AN AIR THAT KILLS

John Rieuwerts obtained a PhD in Environmental Geochemistry from the University of London. He has published numerous academic papers in the field of pollution for more than twenty years and is the author of a university textbook, *The Elements of Environmental Pollution* (Routledge). He currently lives in southwest England and is Senior Lecturer in Environmental Science at the University of Plymouth.

To Gill, Caitlin and Beth

An Air that Kills

Our Invisible Air Pollution Crisis

John Rieuwerts

Also by John Rieuwerts

The Elements of Environmental Pollution (Routledge 2015)

First published in paperback 2016

ISBN 978–1–53–900508–7 (Pbk.)

10 9 8 7 6 5 4 3 2 1

Typeset in 14/16 pt Palatino Linotype by Jay Design.

Contents

1. A Breath of Fresh Air? 1

2. Aer and Smoak: A Smoggy History 6

3. Air Pollution and its Effects 12

4. The Scale of the Problem 26

5. Current Challenges 37

6. Clearing the Air 56

7. Conclusion: Looking Forward to Clean Air for All 85

Epilogue 88

Appendices 91
Notes 95
References 105
Recommended Websites 115
Index 117

CHAPTER ONE

A Breath of Fresh Air?

Get out in the fresh air and walk briskly - and don't forget to wear a smile while you're at it.

Douglas Fairbanks

Most of us probably take clean air for granted and feel that getting out for a 'breath of fresh air' is bound to be good for our health. Sadly, Britain's air is not as fresh as we'd like to think. Air pollution is an invisible killer – it is far more severe than most people realise and the main sources are all around us. All parts of the country can be affected; the remotest corners of the countryside as well as our towns and cities. But, strangely, the scale of the problem is not widely realised. Ask an average British person what they know about air pollution and they will probably talk about greenhouse gases and global warming, perhaps entering a debate about whether climate change is already happening or not. Yet, our routine inhalation of air pollutants is having widespread and deadly serious impacts on the UK population here and now. Smoggy cities are perhaps thought to be part of our past, found today only in hot climates or newly industrialised countries but this is not correct; every year, tens of thousands of deaths in Britain are attributed by medical experts and health professionals to air pollution[1]. Poor air quality is a public health crisis that potentially affects everyone yet is largely unknown or ignored.

Air pollution is not restricted to Britain of course and, on a global scale, poor air quality is linked to *millions* of deaths every year (WHO, 2014a)[2]. In the UK, as elsewhere, those vulnerable include not only the very young and the elderly but also people of any age who suffer from respiratory disorders – the very disorders that can be caused by long-term exposure to air pollutants in the first place. More surprisingly perhaps, the fit and healthy are also at risk from increased exposure, not least because activities like jogging and cycling dramatically increase the volume of polluted air that is inhaled deep into the lungs. And because everyone is involuntarily exposed to air pollutants we are all at some risk of potentially serious health effects. The British government's advisory Committee on the Medical Effects of Air Pollution states that the average reduction in life expectancy of a British citizen caused by unavoidable exposure to air pollution "is larger than that of several other established mortality risks", including road traffic accidents and passive smoking (COMEAP, 2010).

So, why do developed, post-industrial countries like the UK still have dangerously high levels of air pollution, decades after we saw the last of the famously dense smogs in London and elsewhere? The reasons are fairly simple but are not widely known.

Clean Air

Going back to the idea of 'fresh air', let's consider first what we mean by it, so that we can appreciate what is at risk from our polluting activities. Seen from space, the atmosphere is a brilliant but slender halo separating the Earth from the vast blackness of space. It is incredibly precious to us; we live our entire lives in this ocean of air and with every breath we take, we draw its contents deep into our bodies. Most of what we inhale is just two gases, nitrogen and oxygen; they comprise virtually the whole atmosphere and nearly all of the remaining one percent is the less-familiar, noble gas, argon[3]. The remaining, minute fraction contains all the other gases that are naturally present in the atmosphere, from the well-known carbon dioxide and ozone, to the less familiar, such as krypton and xenon. Suspended in the atmosphere are also countless particles. Some of the larger motes of dust are visible to the naked eye but most particles are invisible. Even in clean air, we constantly inhale millions of tiny particles that originate naturally from such exotic sources as volcanoes, deserts and even sea spray.

Nitrogen has dominated the atmosphere throughout the Earth's history and there is still nearly four times as much nitrogen in the air as there is oxygen. In fact, the early atmosphere had none of the life-giving oxygen we are used to today and aerobic organisms like us would suffocate if we time-travelled back to the young Planet Earth. Oxygen only appeared after photosynthesising organisms had evolved in the oceans and started to produce the gas as a waste product. For a long period, this oxygen was mostly used up by reaction with iron in seawater and only after this did surplus oxygen start leaking into the atmosphere. Over time, oxygen (O_2), and the closely related gas ozone (O_3), accumulated in the air, absorbing deadly short wave solar radiation. This allowed evolving life-forms to emerge onto land: previously organisms had been confined to the sea, which also offers protection from the strongest solar radiation[4]. Today, the abundant oxygen in the atmosphere keeps aerobic organisms, including humans, alive.

Although oxygen is the life-giving part of the atmosphere, the air around us and in our lungs is mostly nitrogen. However, being highly stable and unreactive, atmospheric nitrogen has no effect on us, good or bad; we just continually breathe it in and out. The nitrogen contained in every cell of our bodies and every strand of our DNA has to be provided via the food we eat, having first been 'fixed' from the air by invisible microorganisms living in the soil or in colonies on the roots of leguminous plants like peas and clover[5]. It is ironic that we live our entire lives in an ocean of nitrogen but the fraction we inhale is of no direct use to us.

Polluted Air

Pollution of the atmosphere relates to increased levels of 'trace' gases and aerosols that are normally present, in unpolluted air, in relatively minute quantities. Air pollution occurs when the concentrations of one or more of these trace components increases to levels that can cause harm, or in some cases, when synthetic gases or particles are released into the atmosphere.

Air pollution can occur without any human interference at all – air that is uncontaminated by human activities is not necessarily 'clean'[6]. Volcanoes in particular can eject enormous amounts of gases and particles into the atmosphere. As recently as 2010, the Eyjafjallajökul volcano in Iceland erupted, sending 250 million m^3 of ash into European skies. The ensuing rise in atmospheric aerosols led to flight cancellations, economic damage and ruined holidays. The eruption of Mount Pinatubo, Philippines, in 1991 injected millions of tons of trace gases and particles into the global atmosphere; the particles decreased the amount of sunlight received at the Earth's surface, reducing temperatures worldwide. Back in 1883, the even greater eruption of Krakatoa, in Indonesia, spread atmospheric aerosols around the globe, giving rise to spectacular sunsets. More commonly, Britain periodically experiences Saharan dust brought in on southerly winds. During an episode in 2014, Saharan dust created a very noticeable haze in many areas but the dust is more commonly observed after it has settled out of the air, covering cars and other surfaces with a pale orange patina. However, volcanoes, sandstorms and other natural sources of air pollution are the exception rather than the rule and human activities are at the root of most air quality problems.

Ever since the Industrial Revolution, the Earth's atmosphere has been discernibly tainted by anthropogenic pollutant gases and particles, by-products of our industry, fuel combustion and other activities. And we know the atmosphere was carrying contaminants around the world even before this. A remarkable example is the clear signal of ancient Greek and Roman metal smelting that now shows up as traces of lead in deep polar ice cores, corresponding to aerial deposition up to 2500 years ago (Hong et al., 1994). However, it was with the onset of widespread industry and subsequent urbanisation in the eighteenth century that airborne contaminants started to become problematic and the era of gross air pollution began. There are numerous gases and aerosols in polluted air but a smaller number of relatively common atmospheric pollutants have long been of concern because of their serious health effects, coupled with their widespread nature. Most of these key pollutants are gases, in particular ozone and the oxides of both nitrogen (NOx) and sulfur (SOx); the other main category of problematic air pollutants is fine particulate matter (PM), mostly from combustion processes.

In Britain, as in other parts of Europe and North America, industry has declined markedly but we now face a new set of serious air pollution challenges, mainly associated with our extensive use of road vehicles but added to significantly by widespread emissions from modern farming practices. Meanwhile, grossly polluting industries are still present in Russia and 'newly industrialised countries' like India, China and Brazil[7]. So-called 'developing countries' are by no means immune to air pollution either, with sources ranging from the burning of rainforests for agricultural land to the continuing growth of overpopulated and congested cities, often with little regulation. Images of Chinese people wearing smog masks on their way to work are alarming, as are the vast clouds of smoke engulfing the Asian cities downwind of burning rainforests. These events are newsworthy and regularly covered, quite rightly, by British media. However, they may lead some to believe that air pollution is a serious issue only in these faraway places and that in modern Britain we usually enjoy fresh, unpolluted air. Sadly, this is categorically wrong; air pollution in the UK is a largely unseen and ignored phenomenon that is damaging our health and causing mortality equivalent to tens of thousands of British deaths every single year.

CHAPTER TWO

Aer and Smoak: A Smoggy History[1]

Smoke lowering down from chimney-pots, making a soft black drizzle with flakes of soot in it as big as full-grown snowflakes – gone into mourning, one might imagine, for the death of the sun.

Charles Dickens, *Bleak House*, 1853

The poor air quality we experience today continues a long history of air pollution problems, which were first documented in Ancient Greece and Rome. The earliest records in Britain relate to the burning of coal in Nottingham in 1257, when Queen Eleanor, wife of King Henry III, cut short a visit to the city's castle to avoid breathing coal smoke. In the 1280s, commissions were set up to investigate smoke pollution in London and in 1306, a proclamation was issued, apparently by King Edward I, prohibiting the burning of coal in furnaces; the main culprits appear to have been those operating lime kilns but complaints were also lodged against blacksmiths. Pollution from coal burning had grown in London in mediaeval times because traditional wood fuel had become scarcer on the outskirts of the ever-expanding city; however, this situation changed somewhat when the plagues of the fourteenth century reduced population pressures and, therefore, the associated fuel demand.

It was not until the Tudor period 200 years later that population had recovered and once again, complaints about air pollution resurfaced, not least from Elizabeth I, who is said to have ordered a ban on coal burning in London whenever Parliament was sitting. In 1603, Hugh Platt, in his book *A Fire of Coal-Balles,* lamented the damage to plants and buildings caused by air pollution and in 1661, the diarist John Evelyn was moved to publish his '*Fumifugium or The Inconveniencie of the Aer and Smoak of London Dissipated*'. Evelyn bemoaned London's 'horrid' and 'pernicious smoake' and lamented that its skies were 'Ecclipsed with such a Cloud of Sulphure, as the sun it self [...] is hardly able to penetrate'. 'The weary Traveller', he states, 'at many Miles distance, sooner smells then sees the City to which he repairs'. Evelyn made various recommendations to improve the city's air quality but none of them seem to have been adopted, despite the patronage of King Charles II.

With the development of steam power and the onset of the Industrial Revolution in eighteenth century Britain, the atmosphere in cities across the country became thick with a mix of grey smoke and choking gases emitted from the chimneys of both factories and houses. Throughout Britain, street names such as Foundry Lane, Steel Road and Smelter Wood Avenue evoke a smoggy history and whole districts have similar etymologies; in particular, The Black Country, once the industrial powerhouse of the Midlands, is thought by many to have got its name from the black smoke and soot that pervaded everything for miles around[2]. In another example, the town of Hayle in Cornwall, which was an important centre of industry, is based on two separate districts known as Copperhouse and Foundry. More colloquially, London is well known as 'The Smoke'. Edinburgh's age-old nickname of 'Auld Reekie' may well have stemmed from the unsavoury odours of sewage and rotting food that were part of everyday life before improvements in sanitation, although the word reek is said to derive originally from the Old English word *rēc,* meaning smoke, synonymous with the German 'rauch'.

The smogs of the Industrial Revolution are well documented, not only in historical sources, but also in popular culture, most typically in London[3]. The pollution of nineteenth century England is vividly described by Dickens, most memorably in the opening passage of Bleak House (quoted in part at the beginning of this chapter). In *Strange Case of Dr Jekyll and Mr Hyde* (1886), Robert Louis Stevenson describes London fogs 'as brown as umber' and describes 'a great chocolate-coloured pall lowered over heaven." Following Dickens's example, Frances Hodgson Burnett's children's story *The Little Princess* (1905) opens with:

> *"Once on a dark winter's day, when the yellow fog hung so thick and heavy in the streets of London that the lamps were lighted and the shop windows blazed with gas as they do at night…"*

The dense 'pea-souper' smogs, which were known in the capital as 'London Particulars' are often portrayed in films set in Victorian times featuring Jack the Ripper, Sherlock Holmes and the like. Smogs also have a place in art history. The paintings of JMW Turner brilliantly depict the swirling smoke and fogs of Victorian England and many of Claude Monet's Westminster paintings clearly show the yellow-brown skies over London at the turn of the twentieth century. A contemporary of Monet's, the artist William Blake Richmond, complained about the effects of smogs on London's light levels and, in 1898, founded one of the world's first environmental pressure groups, the Coal Smoke Abatement Society. This would later become the National Society for Clean Air and exists today as the charity, Environmental Protection UK.

Concerns about air pollution led to a number of Acts of Parliament between the mid-nineteenth and early-twentieth centuries but legislation was not always successful and some offenders were quick to spot loopholes. For example, the 1875 Public Health Act legislated to control 'black smoke' but at least one polluter gained an exemption when they argued that *their* smoke was dark brown

(Brimblecombe, 1987). As an indication of the slow pace of progress in this era, this particular loophole was not rectified until the 1926 Public Health Act. This Act introduced the concept of pollution control by the 'best practicable means' and this term would inevitably have been used by some businesses as a licence to continue polluting. Victorian business owners and their families may not have been much troubled by the pollution their factories generated in any case if they chose to live (or build their villas) on the clean side of town. Many of our formerly-industrial cities have well-heeled areas of Victorian housing on their south or west side; is this coincidence or might it be because they were upwind of (and therefore unaffected by) the city's pollution? Meanwhile, poorer districts often lie to the east of town, being downwind of the UK's prevailing south-westerlies and presumably copping for most of the smoke in past times. We should remember though that, regardless of the efficacy of legal curbs on industrial smoke, or the historical distribution of housing, there was also a major non-industrial source of air pollution in our towns and villages – the humble domestic fire.

Decades of chronic pollution and deadly smogs in Britain culminated in the 'Great Smog' of December 1952 when thousands of Londoners died from the inhalation of smoke and gases, mainly from domestic fires, that became trapped at ground level for days on end by a mid-winter 'temperature inversion'[4]. Official records at the time stated that four thousand deaths had been caused by the smog; those affected were not just the elderly and infirm but included infants and younger adults. Subsequent interrogation of the contemporary medical records suggests the smog actually caused a further eight thousand fatalities in the following weeks, deaths that were originally attributed to influenza (Bell and Davis, 2001). This was just the latest, and certainly the deadliest, of a number of fatal smogs that had occurred over the previous century[5].

The sheer scale of the 1952 tragedy led to the Clean Air Act of 1956, and further Clean Air Acts in the following years, that allowed local authorities to declare smoke control areas ('smokeless zones'). Gradually, homes became increasingly dependent on central heating and gas or electric heaters, reducing the need for open fires. However, industry, still a major source of air pollution at the time, was given seven years to comply with the Act and, as might be expected, results were not immediate; severe smogs were still being recorded into the 1960s. However, by the middle of that decade, the controls had been largely successful in ridding the UK's most badly-affected urban landscapes of their smoking chimney-pots and 'pea-soupers'. Left behind for future years though, were reminders of Britain's smoky past; in particular, the blackened buildings of cities like Sheffield[6]. Not until the regenerations of the 1980s and 1990s were the sooty coatings of such buildings sand-blasted away.

Such visible pollution had an intriguing biological effect in industrial Britain: the rise of the dark version of the Peppered Moth (*Biston betularia carbonaria*) in the country's parks and gardens. Previously, the lighter coloured type, *Biston betularia typica,* had been dominant; its light-colouring kept it well-hidden from predators when it was resting on lichen covered trees. However, it became much more visible to predators in industrial areas where lichens, which are sensitive to air pollutants, died off and exposed the relatively dark bark beneath. The trees' bark was also darkened by soot deposits in the most polluted areas, further highlighting the presence of *typica*. In these areas, the previously rare *carbonaria* was much less conspicuous against the darker trees and gradually became dominant, as it was missed by predators; however, it has declined again in post-industrial Britain (Cook and Turner, 2008). The case of *Biston betularia* has often been used in schools for demonstrating how natural selection can occur.

The decline of heavy industry from the 1980s onwards further reduced traditional sources of urban air pollutants but, in the last decades of the twentieth century, a different kind of atmospheric pollution, acid rain, came to national and international attention as the sulfur-laden fumes from domestic coal burning were simply transferred from low-lying, house-top chimneys to the tall stacks of power stations that had been newly built outside towns and cities. These new high-level emissions led to long-range and transboundary air pollution as the sulfurous gases and aerosols travelled aloft on high winds, usually across the North Sea to Scandinavia on prevailing south-westerly winds. The acid rain problems of these times were subsequently alleviated to a great extent, largely by emissions control of sulfur and other acidifying substances.

Britain's once industrial might is now the stuff of history books, but Britons' hunger for manufactured goods never disappeared; in fact, our level of material consumption has grown continuously. Today, a large proportion of our goods are manufactured in newly industrialised countries overseas and, as such, Britain has exported most of its pollution whilst continuing to enjoy the products that create it. This is clearly a product of globalisation but, whilst the associated pollution is now a conveniently long way away, it is still being emitted and undoubtedly affecting lives just as precious as our own. Back in Britain, the widespread industrial emissions and smoking chimney pots of the past have declined but, for many years now, a new and insidious source of air pollution has been steadily growing.

CHAPTER THREE

Air Pollution and its Effects

Into my heart an air that kills
From yon far country blows
A.E. Housman, *A Shropshire Lad,* 1896

What Causes Air Pollution in Britain Today?

The visible smogs of previous centuries are much rarer now but the relatively clear skies we see today are deceptive. At times, the air we breathe contains toxic levels of invisible but insidious air pollutants that can be very damaging to our health, regardless of their visibility. But, now that factories and fireplaces no longer fill the skies with smoke, what is the main source of the air pollution that still afflicts Britain? The short answer is road vehicles. Vehicle exhausts emit toxic gases like nitrogen dioxide and benzene but also fine smoke particles that, along with the gases, can reach the deepest and most sensitive passages of the lungs. Car ownership in the UK has risen dramatically in recent decades; numbers of licenced vehicles increased from just under 4 million in 1950 to around 35 million in 2014 (DfT, 2014) and overall motor vehicle traffic volume increased from 29 billion miles in 1949 to 303 billion miles in 2012 (Hobbs and Harriss, 2013). And, for reasons we will come to, more and more of our cars have diesel engines, which are far worse than petrol engines for belching out the very pollutants that cause the worst of today's air quality problems. Britain's busy streets are also home to hundreds of thousands of buses and heavy good vehicles, most of them also emitting diesel fumes that contribute significantly to poor air quality.

Whilst vehicles are certainly culpable for much of our pollution, it would be wrong to ignore other modern-day sources. In particular, wherever fossil fuels are burned, pollution is a likely hazard and our power stations consume huge amounts of fuel every day to provide approximately three-quarters of the country's electricity needs. And, with the theoretical exception of smoke control areas[1], coal and wood are still burned in fireplaces and modern-style wood burners, adding smoke and gaseous pollutants to the skies above villages and other residential areas. Large-scale industry survives too in areas like south Wales and Humberside. In many areas agricultural sources, such as fertiliser applications, animal wastes and biomass burning can be the main causes of poor air quality (Bauer et al., 2016). There are also miscellaneous sources, such as waste incineration and shipping. In addition, transboundary pollution from the European continent, with millions of vehicles of its own, as well as areas of intensive agriculture and industry, adds significantly to the UK's poor air quality.

To put the importance of these various sources into perspective, research at the Massachusetts Institute of Technology indicates that some 39% of the life expectancy lost in Britain to pollution from combustion sources can be attributed to UK transport emissions (mainly road vehicles), 13% to UK power generation, 4% to Britain's remaining industry and 32% from transboundary pollution from Europe, with its similar mix of emission types; the remainder is attributed to miscellaneous combustion sources, including residential and agricultural activities[2] (Yim and Barrett, 2012).

Air Pollutants of Particular Concern

This mix of pollution sources, dominated as it is by vehicle emissions, puts vast quantities of gases and particles into our skies each day; but there are some pollutants that are of particular concern because of their sheer quantity combined with their toxic effects. As such, the main culprits are fine particles and the gases nitrogen dioxide (NO_2) and ozone (O_3), whilst other potentially problematic pollutants include sulphur dioxide (SO_2), carbon monoxide (CO), various hydrocarbon gases and toxic trace elements. For example, the European Union (EU)[3] sets legally-binding 'limit values' (maximum concentrations) for seven air pollutants and, for five others, less stringent 'target values', concentrations to be attained where possible, 'not entailing disproportionate costs' (Table 3.1). The limit values are implemented mainly in an attempt to limit human health impacts from toxic levels of these key pollutants[4].

Table 3.1. Key air pollutants of concern.

Pollutants with EU *limit* values	Pollutants with EU *target* values
$PM_{2.5}$[a]	Ozone (O_3)
PM_{101}[a]	Polycyclic aromatic hydrocarbons
Nitrogen dioxide (NO_2)	Arsenic
Sulfur dioxide (SO_2)	Cadmium
Carbon monoxide (CO)	Nickel
Benzene	
Lead	

a. For an explanation of the terms $PM_{2.5}$ and PM_{10}, see text (page 15).

Particulate Matter

The excess particulate matter (PM) measured in polluted areas is derived mainly from combustion sources. Larger particles, such as those comprising a cloud of smoke are clearly visible to the naked eye. The smallest particles, which raise the most health concerns, are contained not only within thick smoke but also in less visible exhaust emissions, such as those from modern vehicles. The highest levels of particulate pollution occur today mainly in the vicinity of busy roads in urban areas, although pockets of remaining heavy industry can also be significant local sources. Vehicles produce fine PM from the combustion of fossil fuels, especially diesel, but tyre and brake wear from all vehicle types also generate a significant amount of fine particles, as does road abrasion, and these non-exhaust sources are accounting for an increasing proportion of urban PM pollution as exhaust emissions are gradually reduced by the use of particulate filters. Importantly, a significant amount of pollutant PM is secondary in nature, meaning that, rather than being emitted directly from exhaust pipes and chimneys (i.e. 'primary' PM), it is *produced in the atmosphere* from chemical reactions of pollutant gases, particularly nitrogen dioxide, sulfur dioxide, ammonia and volatile organic compounds; for this reason, agriculture is a major cause of PM pollution as it is such a widespread source of the precursor gases[5]. Further PM sources include domestic heating, waste incineration and dusty activities like construction and quarrying. More rarely, natural PM sources can add to the pollution mix, particularly when Saharan dust is brought to the UK by southerly winds.

For a long time, the classification of fine PM was based entirely on 'PM_{10}' particles – those with a diameter of less than 10 micrometres (a micrometre, µm, is a thousandth of a millimetre). The PM_{10} category has long been used in the investigation and management of air pollution and subsequently it has been joined in this regard by a class of even smaller, 'ultra-fine' particles known as $PM_{2.5}$, which relates to particles less than 2.5 µm in diameter. Both types of PM are far too small to be seen with the naked eye.

Fine and ultra-fine PM are of greater interest than the coarser and more visible particulate matter because they are small enough to travel deep into the lungs, where they can have harmful effects[6]. Particles that are larger than 10 µm tend to be deposited in the throat and coughed up in phlegm but PM_{10} is transported deep into the bronchi and bronchioles of the lungs and $PM_{2.5}$ can travel all the way down to the alveoli, the tiny air sacs at the ends of the finest lung passages. The alveoli are where oxygen is passed into the bloodstream via hundreds of tiny capillaries that cover each air sac. However, it is thought that smaller $PM_{2.5}$ particles may also pass through the alveoli walls into the bloodstream, in some cases carrying toxic metals and other harmful pollutants, such as carcinogenic polycyclic aromatic hydrocarbons (PAHs) (WHO, 2012). There is also evidence to suggest that ultra-fine PM is deposited in the brain, where it could have direct toxic effects (Maher et al., 2016).

It is well established that short- and long-term exposure to pollutant levels of PM_{10} and $PM_{2.5}$ can have serious health effects. It causes chronic obstructive pulmonary disease (COPD), which encompasses a number of respiratory diseases including emphysema and bronchitis. Particulate pollution is also associated with increased incidence of cardiovascular disease (affecting the heart and blood vessels) including stroke and heart attack. These effects lead not only to increased incidence of ill-health and hospital admissions but also death in severe cases. Some of the key symptoms of PM exposure are asthma, wheezing and reduced lung function; it also affects lung development in children. Exposure to PM is also linked to increased incidence of lung cancer. Chronic exposure to $PM_{2.5}$ is thought to cause a 6-13% increase in the long-term risk of death from cardiopulmonary diseases for every 10 $\mu g\ m^{-3}$ increase in ambient $PM_{2.5}$ concentration (WHO, 2013). Estimates of excess mortality related to PM exposure will be detailed in Chapter 4.

Nitrogen Dioxide

Among the gaseous pollutants, nitrogen dioxide (NO_2) is particularly troublesome, not only because of its direct health effects when inhaled but also because it can react in the atmosphere to create two other harmful pollutants, the noxious gas ozone (see next section) and ammonium nitrate, which is an important fraction of $PM_{2.5}$[5].

Nitrogen dioxide is formed when fossil fuels are burned but there are two distinct mechanisms that generate NO_2 from fuel combustion. Firstly, fossil fuels contain the element nitrogen, a relict of the nitrogen content of the organisms that were transformed into coal and oil millions of years ago; when the fuel burns, the fossilised nitrogen reacts with oxygen in the air to form NO_2. Secondly, and quite separately, when any combustion process heats the air to a sufficiently high temperature, the two most common components of the atmosphere, nitrogen (N_2) and oxygen (O_2), can be fused together as NO_2[7]. The two gases do not react under normal conditions, because atmospheric nitrogen is a particularly unreactive molecule; this is because there is a very strong triple bond between its two nitrogen atoms that can only be broken with a large energy input. When an air mass is heated sufficiently (by high temperature combustion processes, or naturally by lighting bolts) the individual nitrogen atoms can be liberated from the N_2 bond and go on to react with individual oxygen atoms in the air, which are more easily liberated from the gaseous oxygen (O_2) molecule.

For the reasons outlined above, NO_2 can become a pollution problem wherever fossil fuel is burned, in vehicle engines, industrial processes, power stations and other combustion sources. Nitrogen dioxide pollution is widespread across the UK, especially in congested urban areas and in the proximity of main roads. On colour-coded maps of NO_2 pollution across the UK, produced by the government's Department of Environment, Food and Rural Affairs, major towns and cities can easily be picked out, as can motorways, which show up as thin strands of pollution running across the countryside (DEFRA, 2015a). In 2013, roadside monitoring across the UK indicated that limit values for NO_2 were breached along some 2,500 km of British roads (DEFRA, 2015b).

According to the World Health Organisation (WHO), short-term exposure (e.g. for a period of one day) to NO_2 concentrations exceeding 200 $\mu g\ m^{-3}$ can severely inflame the airways, causing bronchial constriction and breathing difficulties, particularly in asthmatics (WHO, 2000, 2014b). Over longer exposure periods (e.g. one year) studies tentatively suggest an association between concentrations exceeding 50 $\mu g\ m^{-3}$ and a reduction in the growth of lung function in children (WHO, 2000). Existing respiratory conditions are also aggravated by exposure to elevated levels of NO_2. It has been shown in studies that short-term exposure can reduce the ability of the respiratory system to clear pathogens, allergens and fine PM pollutants from the lungs (a process called mucociliary clearance) although this appears to happen only after exposure to levels that are considerably higher than those normally encountered in ambient air. As well as the direct health impacts of elevated NO_2, its role as a precursor of fine PM (above) and tropospheric ozone (below) needs also to be borne in mind as these pollutants also have a direct impact on human health. Health experts believe that NO_2 exposure in the UK is a major contributor to excess mortality (as detailed in Chapter 4).

Ozone

When ozone is mentioned, most people tend to think of the ozone 'layer' and its intermittent 'hole'; however, this issue relates to high-altitude ozone in the stratosphere and has nothing to do with the serious health problems that can be caused by ozone inhalation at ground level. Whilst there is actually no such thing as a distinct, single ozone layer in the stratosphere (the gas is present at all levels of the atmosphere[8]), ground-level ozone concentrations are certainly much lower than they are in the stratosphere. The trace concentrations of ground-level ozone can increase dramatically however during a 'photochemical smog', when two factors coincide; the presence of NO_2 pollution and sunny weather. In a smog of this type, sunlight catalyses chemical reactions between nitric oxide (NO), NO_2 and various volatile organic compounds, all of which are emitted by vehicle exhausts and other combustion sources. These 'primary' pollutants yield ozone and other 'secondary' pollutants (i.e. pollutants that are not actually emitted from a primary source). In other words, despite being a serious pollutant, ozone is not emitted by vehicles or any other source but is produced in sunny skies that have been polluted by other gases.

An anomaly of the secondary nature of ozone is that by the time it has been produced in quantity within a mass of polluted air, the air mass and its growing ozone concentrations may have travelled a long way from the original primary pollutant sources; not just into rural areas but also across national borders. In recognition of the transboundary nature of much ozone pollution, the EU does not stipulate a legally-binding limit value for ozone, despite its serious health effects, acknowledging that measures taken by an individual member state cannot always guarantee that a limit value will be met; i.e. if that country is receiving imports of ozone from abroad. Ozone is, unusually, a problem pollutant in rural areas and is typically present at higher concentrations in the countryside than in urban areas. This is partly because of the long-range transport described but also because a significant fraction of ozone in urban areas is destroyed by reaction with fresh nitric oxide emissions from vehicles. The presence of pollutant ozone in the countryside may come as a surprise to many but it clearly highlights the fact that health-damaging pollution is not restricted to those living in our congested towns and cities.

Inhalation of excess ozone over a period of a few hours can cause serious respiratory problems. The main symptoms of both short- and long-term exposure are: throat irritation, coughing, wheezing and shortness of breath, chest tightness and burning pain when inhaling deeply. These are symptoms of the very damaging effects that ozone can have on the respiratory system. Unlike some other pollutant gases, ozone has limited water-solubility, so it is not as effectively removed from inhaled air by the moist linings of the mouth, nasal passages and throat, leaving more to be taken into the lungs. Worse still, ozone is a highly reactive gas and reacts with protein and fat molecules that are an integral part of the inner lining of the lung passages. Ozone and other oxidants associated with these reactions can directly damage the cells lining the airways as well as other specialised structures in the lung, such as immune cells. Severe inflammation of the airways can result from the damage ozone causes and lung function is decreased in affected subjects after short-term exposures of a few hours. Permanent scarring of the lungs can result from long-term exposure. The lung damage caused by ozone also increases vulnerability to lower respiratory tract infections and sensitivity to allergens.

The WHO warns that children may suffer disproportionately from ozone inhalation in polluted environments because on average, compared to adults, they do more physical activity, have a higher metabolic rate and spend greater amounts of time outdoors. On the other hand, health statistics on elderly people show a pronounced association between ozone exposure and early death. But ozone inhalation does not only affect the youngest and oldest in society; ozone exposure is a particular health risk for people exercising heavily in polluted air for more than a short period. This is because, as well as increasing the volume of air inhaled, heavy exercise also causes deeper breathing, meaning that more ozone is taken down into the deepest and most sensitive parts of the lungs. For all of these reasons, the WHO recommends guideline levels for ozone in the air that should not be breached. Unfortunately, as we will see, most parts of the UK, rural as well as urban, regularly experience unsafe levels of ozone, in excess of these guideline concentrations.

Sulfur Dioxide

Just as the nitrogen content of burned fossil fuel is released as NO_2, so the sulfur in coal and oil yields atmospheric SO_2, sulfur dioxide[9]. Despite our continuing dependence on fossil fuels however, SO_2 pollution has decreased markedly since the times of the pea-souper smogs of the past. By the 1970s, the Clean Air Acts of the 1950s and 60s had already made an impression, although annual mean SO_2 concentrations of 150-200 $\mu g\ m^{-3}$ were still being recorded in central London. But by the first decades of the twenty-first century, annual average concentrations in London had fallen to 2-12 $\mu g\ m^{-3}$. The relative success story for SO_2 has been possible because of behavioural and technological changes. First came the declaration of smoke control areas after the Great Smog in 1950s London, which forced a shift away from the domestic burning of sulfurous coal in towns and cities. Secondly, technology has allowed the removal of sulfur from transport fuels and the desulfurisation of industrial waste gases, particularly those from power stations, which were formerly a major source of SO_2. Some residual problems still persist however, not least emissions from shipping, which has traditionally been allowed to use heavy fuel oils. This is now being tackled as well and the SO_2 decline over the last half-century shows how air quality can be improved when the will is there.

Despite the gradual improvements relating to SO_2 pollution there is no room for complacency. The WHO (2014b) states that "health effects are now known to be associated with much lower levels of sulfur dioxide than previously believed". The symptoms of excess SO_2 inhalation include COPD (mainly bronchial constriction), asthma aggravation and decreased lung function, particularly in children. As with nitrogen dioxide, SO_2 is also a $PM_{2.5}$ precursor thereby causing further, indirect health effects.

Other Harmful Air Pollutants

The European Union stipulates legally-binding limit values within its member states for three other air pollutants, carbon monoxide (CO), benzene and lead. The main source of carbon monoxide in polluted areas is vehicle emissions. Inhalation of carbon monoxide can be deadly, as seen all too often in news items about accidental exposure in confined spaces, via faulty heaters for example. Concentrations in the ambient atmosphere are much less than those recorded in such fatal cases but the EU has set a limit value for CO, aiming to protect its populations from exposure to elevated levels at the road side and in partially-enclosed spaces like road tunnels. Benzene is a carcinogen, known from industrial exposure studies to cause leukaemia at high levels of exposure. It is another combustion product but also arises from volatilisation of gasoline, at petrol stations for example. Lead, rather like sulfur dioxide, is less problematic as an air pollutant these days, mainly because of its removal from petrol many years ago, but it can still be detected in fine particulate matter. It is a very toxic element, causing neurological defects and other health problems.

A number of other air pollutants are subject to some level of monitoring in the EU and/or the UK, including the toxic elements arsenic, cadmium and nickel. Others include gaseous hydrocarbons such as benzo(a)pyrene and 1,3-butadiene, both of which are carcinogenic. The latter gas is also known from occupational studies to be mutagenic and to have cardiorespiratory and central nervous system effects. Despite the understandable concerns about such toxic gases however, and the need for continued vigilance, the most widespread pollution in modern-day Britain is undoubtedly caused by $PM_{2.5}$, PM_{10}, NO_2 and tropospheric ozone.

Pollution Episodes

The pollution sources already described are constantly emitting potentially harmful gases and particulates into the atmosphere. In the most polluted areas, near congested roads or industrial plants for example, there are ongoing problems of chronic, long-term air pollution. More generally though, Britain's worst pollution problems tend to come in short-term 'episodes' governed by specific weather conditions that effectively accumulate and concentrate the pollutants at ground level, leading to greatly increased rates of exposure to human populations.

I sometimes ask my students; what is likely to happen to a city's polluted air over a period of a few hours? Usually, the first answer is that it will begin to blow away if there is any wind and that this will dilute the pollution to some extent. This is generally true for Britain as its winds are predominantly south-westerly in direction, bringing in clean air from the Atlantic Ocean. If the wind swings round to the south-east then the incoming air, from the European continent, can import relatively polluted air to large parts of the country but as a general rule, the windier the weather, the more dilution of local pollution will occur. Sometimes, a student may point out that air also rises away from ground level if it is warm enough. This is also correct: except on very cold days, the ground is warmed to some extent by direct or filtered sunlight and this warms the air in contact with it causing it, and any ground-level pollution, to rise away from ground level, just as a hot air balloon gently rises in the atmosphere. This movement of polluted air away from ground level is another form of dilution. Another correct response I sometimes hear in the lecture theatre is that Britain's typically wet weather washes pollutants out of the air; this is true for gases as well as particles. If we have been talking in class about the production of ozone in the sunny conditions typical of a photochemical smog, then students might also point out that cloudy skies will make this kind of pollution less likely in the first place.

All the meteorological conditions described above, which tend to decrease pollution levels (i.e. high winds, ascending air, cloud and rain), are common in periods of low pressure, which have rising air at their centres. These are the 'depressions' we frequently see on weather forecasts that bring in cloudy, wet and windy weather, usually from the Atlantic. Most people are happier when high pressure systems bring in calmer and clearer weather, but in pollution terms 'highs' are bad news; whilst they tend to leave us drier, less wind-blown and, in some cases, sunnier they can also allow pollutants to develop and/or accumulate at ground level, sometimes leading to pollution episodes. So why is this the case?

When high pressure conditions occur in the tropics and sub-tropics (and elsewhere during summer months), not only are wind-driven dilution and rainout of pollutants less likely, but the relatively strong sunlight can drive photochemical reactions of primary pollutants, adding secondary pollutants like ozone and some fractions of $PM_{2.5}$ to the air. Photochemical or 'summer' smogs'[10] of this kind are most common in sunny locations like Los Angeles, where this type of 'smog' was first detailed, but can occur during summer at higher latitudes, such as those of the UK. It might be thought that the relatively warm conditions would at least accelerate the upward dilution of pollutants as described above. In fact, high pressure counters this too; high pressure at ground level is caused by an air mass at altitude pushing downwards towards the Earth's surface and as it does so the air mass is compressed and warmed just like the air being pumped into a bicycle tyre. This creates something called a temperature inversion, where the compressed, warm air sits above *relatively* cool, denser and heavier air at ground-level. This is an 'inversion' because it is the opposite of the typical situation where air temperature decreases with altitude. The problem with temperature inversions, in the context of pollution, is that ground-level air will only rise if it is warmer than the air above it. This is why hot and sunny cities like LA suffer from temperature inversions and summer smogs even though the air in the city could not, in itself, be described as cool; only *cooler* than the air higher up (also, in the case of LA, ground-level air is often cooler maritime air that has blown onshore from the Pacific Ocean). For the same reasons, photochemical pollution episodes sometimes occur in the UK too, on those occasions during summer when the weather becomes warm and sunny.

High pressure conditions can also bring pollution episodes in winter, for some, but not all, of the same reasons. The normal dilution and removal mechanisms (e.g. wind and rain) are limited, as in summer, leading to a heightened risk of pollution. A clear difference with the high pressure conditions in summer though is that photochemical reactions are unlikely to occur during winter in the temperate zones of northern Eurasia and America; this is because the sun, even if it is out, is always quite low in the sky and therefore much weaker than in summer. Therefore, the pollutants associated with summer smogs, particularly secondary ozone, are generally not a problem in winter. Primary pollutants however can accumulate to greater levels than in summer, particularly because temperature inversions in the winter are likely to be at their strongest. This is because the air at ground level, typically in contact with the relatively cold surface, is often colder than the air further up in the atmosphere that is not being chilled in the same way; in high pressure periods, this higher level air is also being warmed by compression as it sinks and comes to rest above the cool ground layer. As in summer then, the colder ground-level air, being relatively dense, cannot rise and dilute the pollutants upwards. The usual pollutants from traffic and industry, together with those from additional winter sources, particularly domestic heating, are released into this still air and start to accumulate in the absence of any dilution and dispersal. The Great Smog of London occurred in just such conditions in December 1952 and the UK still suffers from winter pollution episodes today.

Table 3.2. Causal factors in air pollution episodes

Causes	Winter-type	Summer-type
Additional emissions (heating)	✓	
Temp. inversions (cold surface)	✓	
Temp. inversions (sinking air)	✓	✓
Photochemical reactions		✓
Low winds	✓	✓
Low rainfall	✓	✓

In summary, the UK suffers at times from summer-type and winter-type pollution episodes. At these times, the pollution that is emitted into the atmosphere year round can accumulate at ground level, exposing human populations to particularly unhealthy concentrations of the most serious air pollutants, such as fine PM and NO_2. In summer, sunny skies can, in addition, lead to the generation of secondary pollutants, particularly ozone, which affect not just population centres but also rural areas remote from the original sources of pollution.

CHAPTER FOUR

The Scale of the Problem

> *The number of lives cut short by air pollution is already terrible and the potential rise in the next few decades is terrifying.*
>
> Simon Upton, OECD

The Human Tragedy of Air Pollution

Air pollution in the UK today is responsible for raising rates of lung cancer, causing cardiorespiratory diseases, reducing lung development in children and leading to tens of thousands of attributable deaths every year. On a global scale, the WHO (2014a) estimates that 3.7 million deaths each year can be attributed to ambient air pollution; 80% of these are attributed to heart disease and stroke, 11% to COPD, 6% to lung cancer[1] and the remaining 3% to respiratory infections in children. The deaths are estimated to be particularly concentrated in east and southeast Asia, whilst in the WHO European Region, outdoor air pollution was linked to an estimated 482,000 deaths in 2012 (WHO, 2015). The prognosis for the foreseeable future is even worse. The Organisation for Economic Co-operation and Development (OECD) estimates that annual deaths attributed to air pollution will rise to between 6 million and 9 million globally by 2060 (OECD, 2016).

For the UK, an estimate by Public Health England (Gowers, 2014) of 28,969 'attributable' deaths per year featured in the media, although this estimate relates to only one type of air pollution, $PM_{2.5}$[2]. The additional effects of various other pollutants in Britain's atmosphere are being investigated and it is thought that future estimations of attributable deaths could be much higher, although there is likely to be considerable overlap between the contributions of individual pollutants to mortality[3]. A study by researchers at Kings College London (Walton et al., 2015) indicated that the number of deaths linked to air pollution is likely to be much higher than previously thought because of the effects of other pollutants, particularly NO_2; the study cautiously estimates 9,416 attributable deaths in London alone (for the year 2010) with only 3,537 of these ascribed to $PM_{2.5}$, the pollutant used as the basis for the national mortality figure mentioned above. More recently, DEFRA, the UK government's environment department, has estimated that NO_2 pollution has effects on mortality equivalent to 23,500 deaths in the UK each year (DEFRA, 2015b). Subsequently, The Royal College of Physicians (2016) published a widely-quoted and influential report that considered all the evidence presented previously and estimated that approximately 40,000 UK deaths every year are attributable to the combined effects of PM and NO_2 exposure.

These alarming mortality estimates illustrate that air pollution is a hugely significant health hazard in modern Britain, as well as in the wider world. Yet, it is seldom discussed and most people would probably not name poor air quality as one of the main causes of serious ill-health and death in the UK today. They would perhaps think first of better-known and well-established mortality risks like road accidents and passive smoking, not realising that air pollution is thought to have a greater effect on mortality than both. The UK Committee on the Medical Effects of Air Pollution estimates that ambient air pollution in Britain causes 2.5 to 6 times more loss of life expectancy than motor-vehicle accidents and 2.5 to 3 times more than exposure to environmental tobacco smoke; the lower and higher figures in each case relate to males and females, respectively (COMEAP, 2010). The number of deaths attributed to alcohol consumption in the UK (8,697 in 2014 (ONS, 2016)) is also much lower than the estimated mortality related to air pollution.

Who is at Risk?

The overall public health impact of air pollution is increased by the fact that everyone is at risk of involuntary exposure. And this highlights an element of unfairness inherent in the disease and death caused by air pollution; it is unavoidable and indiscriminate. For example, each year around the world, 127,000 deaths of young children (less than five years old) are attributed to ambient air pollution (WHO, 2014a). Unlike other 'avoidable' causes of death (e.g. overly sedentary lifestyles, over-consumption of calorific foods and alcohol, tobacco smoking, participation in high-risk activities) exposure to air pollution is not directly self-inflicted. We can all try to 'do our bit' to reduce it and could, arguably, do more (see Chapter 6); however, air pollution is largely created by societies at large and individuals have no chance of totally avoiding it, regardless of their own level of concern or contribution.

The risk of ill-health is increased further for anyone living and working in particularly polluted areas such as inner cities or in close proximity to congested roads (Table 4.1). People undertaking strenuous exercise (such as joggers, cyclists and athletes) are also at particular risk of enhanced exposure to air pollutants because of their deeper and heavier breathing rates. This is likely to be most acute in congested urban areas but such exercising could cause problems almost anywhere during pollution episodes. In this context, it is worth remembering that ozone, which is generally at its highest concentrations in rural areas, is a particular hazard to those undertaking heavy exercise. In general, the health benefits of exercise are likely to outweigh mild exposure to pollutants but at certain times and places this may not always be so; and, in any case, those who choose to exercise and look after their health would presumably prefer not to be at enhanced risk of lung damage or other pollution-related disorders.

Table 4.1. Exposure to air pollution: Key locations and groups at particular risk.

Location	Those at greatest risk of heightened exposure to air pollution
Congested urban areas	Inhabitants Workers Regular visitors
Industrial areas	Those living in close proximity Those working in close proximity
Main roads, including those outside urban areas	Those living in close proximity Workers Vehicle occupants (inhaling pollutants emitted from vehicles ahead)
Rural areas	Inhabitants, workers and visitors (inhaling long-range ozone that is typically present at higher levels in the countryside than in urban areas)

Some groups in the population are especially vulnerable to the *effects* of inhalation. As well as the very young, this includes frail, elderly people and those of any age who already have cardiovascular and respiratory conditions such as asthma, bronchitis and emphysema. The types of damage caused by the most common pollutants, as detailed in Chapter 3, are clearly likely to have more serious consequences for such vulnerable people, compared to those who are relatively healthy.

The Scale of Air Pollution in the UK

The levels of air pollution in Britain today can easily be observed by anyone with access to the internet because the UK has an extensive air quality monitoring system called the Automatic Urban and Rural Network (AURN). This nationwide network of more than a hundred monitoring stations, located mainly in cities but with some rural stations too, measures the levels of the main air pollutants around the clock and sends the results to a publicly-accessible database (DEFRA, 2016a). Evaluating whether the reported concentrations pose potential risks to health requires knowledge of guideline levels set by the World Health Organisation. Their published Air Quality Guidelines (WHO-AQG) are based on research into the health effects of air pollutants and are generally used as the basis for the EU's 'limit values', which are translated into UK legislation as 'air quality standards' (AQS). For example, the UK AQS for daily-averaged PM_{10} is 50 μg m^{-3} (i.e. 50 micrograms of PM_{10} in every cubic metre of air); this means that the measured PM_{10} concentrations for a particular location in the UK, when averaged out over a whole day[4] should not exceed 50 μg m^{-3}. The AQS for the key UK air pollutants are shown in Table 4.2[5]. As we shall see later, meeting these guideline concentrations does not necessarily safeguard entirely against health impacts but they are a useful set of benchmarks to judge the extent of Britain's current air pollution.

Table 4.2. The United Kingdom's air quality standards (AQS) for the main air pollutants of public health concern. The standards are set to meet the EU's legally-binding 'limit values'[a], which are in turn based on health research collated by the World Health Organisation.

Pollutant	Measurement	UK AQS (μg m^{-3})
NO_2	Hourly mean	200
NO_2	Annual mean	40
O_3	8-hour mean	100[a]
PM_{10}	Daily mean	50
PM_{10}	Annual mean	40
$PM_{2.5}$	Annual mean	25
SO_2	Hourly mean	350
SO_2	Daily mean	125

a. For O_3, there is no legally-binding EU limit value but a non-mandatory target value of 120 μg m^{-3} (see Chapter 3).

Exceedances of the Air Quality Standards

It might be assumed that pollutant levels in a post-industrial country like the UK rarely, if ever, exceed these concentrations that have been set as standards to protect health. In reality, the main pollutants of concern regularly breach the health standards in the UK year after year. Table 4.3 summarises the extent to which the UK's air quality standards have been exceeded in recent years, according to data from the national AURN monitoring programme. Exceedances of PM_{10} and O_3 occur at the vast majority of monitoring sites across the entire country.

Table 4.3. Percentage of official UK monitoring sites exceeding UK Air Quality Standards, 2010-15. Based on raw data from DEFRA (2016b). See Appendix 1 for a detailed breakdown.

PM_{10} (daily mean)	NO_2 (hourly mean)	O_3 (8-hour mean)
84-95%	13-18%	89-95%

During the period shown in the table, the daily AQS for PM_{10} of 50 μg m^{-3} was exceeded (on one or more days each year) at the vast majority of the sixty-plus UK sites that monitor for it. None of the sites exceeded the lower *annual* AQS for PM_{10} of 40 μg m^{-3} (not shown in the table) although, as we will see, this does not necessarily mean that long-term PM_{10} exposure is not a problem in the UK. The large percentages of UK sites experiencing daily exceedances gives cause for concern about short term exposure, particularly in those specific locations (not shown) where the daily AQS is breached repeatedly throughout a year. Taking 2014 as an example, all but ten of the UK's sixty-five PM_{10} monitoring sites exceeded the AQS during the year. Most exceeded the AQS on five or more days during the year and twelve sites breached the limits on ten or more days; the worst offender was the London Borough of Ealing with exceedances of the AQS on twenty-one separate days during the year. Similarly, in 2015 most of the stations monitoring for $PM_{2.5}$ (fifty-four out of a total of seventy) recorded annual mean concentrations at or above the WHO's guideline level of 10 μg m^{-3}.

The extent of exceedances measured for NO_2 by the AURN in recent years is lower than for PM_{10}; however, this smaller percentage range includes some sites, notably in large cities such as London, Birmingham, Glasgow and Leeds that consistently suffer from high NO_2 levels. For example, every year between 2011 and 2015, the average annual concentration at London's Marylebone Road has been more than twice the annual AQS for NO_2 of 40 $\mu g\ m^{-3}$ and maximum hourly concentrations, which were recorded in Belfast, London and Dumfries, were 339-462 $\mu g\ m^{-3}$ (the hourly AQS for NO_2 is 200 $\mu g\ m^{-3}$). Additional monitoring in London, by the Environmental Research Group at Kings College, provides data showing that other sites in London have even higher NO_2 concentrations; for example, hourly concentrations at Oxford Street exceeded 600 $\mu g\ m^{-3}$ in 2015 (ERGKC, 2016a). Nitrogen dioxide is also monitored widely by local authorities at roadside sites and NO_2 pollution is known from that extensive network to be a very widespread problem in the UK. Of the 389 local councils in the UK, 248 have declared 'air quality management areas' (AQMAs) for one or more locations in their local authority area, meaning that action plans are required to investigate how elevated levels of air pollution at those locations might possibly be addressed; between them, these 248 councils have declared 579 individual AQMAs for breaches of NO_2.

The percentages of AURN monitoring stations exceeding the AQS for ozone (89-95%) are as high as those for PM_{10} (Table 4.3). But there are two key differences. First, the *numbers* of exceedances per year at *each* of the monitoring locations are higher than for PM_{10}. To give an idea of this, in 2014 there were, on average, thirty-seven exceedances per monitoring site across the AURN network. This indicates that ozone pollution across the country can be persistent and/or recurring as well as widespread. The other difference is that, uniquely for ozone, many of the AQS breaches occur at rural sites. In 2014, the daily maximum concentrations for rural ozone were nearly 10 $\mu g\ m^{-3}$ higher than for urban ozone, when averaged across all sites for the year, continuing a long-term trend going back many years.

The evidence base on air quality that is provided by AURN monitoring is supplemented by national modelling of air pollution, based on: (i) background concentrations on a national 1x1 km grid (using detailed information on pollution sources) and; (ii) roadside concentrations at major roads throughout the UK (using information including road traffic counts). This modelling is valuable as it gives an indication of air pollutant concentrations across the whole country, not just in the immediate vicinity of the 100-plus AURN monitoring stations. The model is validated by comparison of the modelled concentrations against measurements from the AURN and other monitoring networks. DEFRA uses the modelled and measured concentrations to determine the extent of compliance of the UK's air quality with EU standards and objectives and this is reported for forty-three regional zones covering the entire country. The results of the modelling add to the evidence we have seen from the AURN that air pollution is severe and widespread across the whole UK. In 2014 for example, all except five of the regional zones were reported to be non-compliant with the annual limit value for NO_2; for O_3, thirty-two zones exceeded the long-term EU objective (DEFRA, 2015a).

Pollution Episodes in the UK

The figures in Table 4.3, and the national modelling outputs, indicate just how widespread air pollution is in the UK and we have seen that it can affect all parts of the country. In our towns and cities, but also in the countryside, UK citizens are exposed to potentially-harmful levels of air pollutants. Some areas regularly suffer from higher levels of pollution than others but, as explained in Chapter 3, an added concern is that certain weather conditions can increase the risk of particularly severe air pollution occurring during 'pollution episodes', not just in the normally-affected areas but in others too.

There are frequent pollution episodes in the UK. On some occasions they are caused by still, dry and, often, cold conditions limiting the sideward and/or upward dispersal of pollutants and, on others, by sunny weather generating photochemical smogs. Imports of air pollution from European countries add to the problems when the UK's winds swing around from the usual Atlantic south-westerlies (a source of clean air) to continental southerlies and easterlies. Table 4.4 lists pollution episodes recorded throughout 2015 in southeast England and serves to indicate how common such episodes are and the various reasons for their occurrence in the UK.

For the UK as a whole, DEFRA annually reports the main pollution episodes for the previous year in its 'Air Pollution in the UK' reports, available on its website. Table 4.5 shows a summary of the episodes detailed in the report for 2013. It is instantly clear from Tables 4.4 and 4.5 that PM (and NO_2) pollution is associated with colder conditions at the beginning and end of the year, whilst ozone, a photochemical pollutant, is more problematic in spring and summer when the sun is high and strong enough to drive reactions of primary pollutants. Table 4.5 shows that typical British weather conditions during 2013 led to 109 days when pollution levels breached health guideline levels over wide areas of the country, not just in the more usual 'hotspots' in busy city centres. This was not an exceptional year for weather or pollution emissions and the long list of episodes gives an indication of the scale of air pollution across the UK.

Table 4.4. Pollution episodes in London and southeast England, 2015. Based on information accessed at ERGKC (2016b)[6].

Month	Conditions	Pollutants	Causes
January	Cold and calm	$PM_{2.5}$ PM_{10} NO_2	Local accumulations of the primary pollutants PM and NO_2 in conditions not conducive to their dispersal; some E winds later in the episode also imported industrial pollution from Europe.
February	Cold and calm	PM_{10} NO_2	Still conditions leading to local (near-source) accumulations of PM and NO_2.
March	Light E and SE winds Later, still conditions	$PM_{2.5}$ PM_{10}	$PM_{2.5}$ imported from Europe, attributed mainly to springtime fertiliser applications on the continent generating ultra-fine nitrate and ammonium particles. Later in the month, still air led to locally elevated PM_{10} levels.
April	Calm	$PM_{2.5/10}$ NO_2	Low winds leading to accumulation of primary pollutants.
June	Sunny and SE winds	O_3	Photochemical reactions of primary pollutants emitted in the UK and Europe leading to elevated ozone concentrations.
August	Sunny and hot	O_3	Unbroken sunshine leading to elevated ozone across London and Sussex
October (early)	Calm; continental air	$PM_{2.5}$ PM_{10}	Still air led to poor dispersion of PM from local and imported pollution at roadside and urban background locations
October (mid/late)	Calm	NO_2	Still air led to poor dispersion of NO_2 at roadside locations
November	Calm and foggy	$PM_{2.5}$ PM_{10}	Accumulation of PM at roadsides and in residential areas. Contribution from fireworks and bonfires possible.
December (mid)	S winds	PM_{10}	Imports of Saharan dust across SE England added to local traffic and industrial sources
December (late)	S winds	PM_{10}	Imports of Saharan dust across SE England added to pollution from domestic wood-burners (little traffic or industry input)

Table 4.5. UK pollution episodes relating to particulate matter (PM) and ozone (O_3) in a typical year (2013). Based on information in: DEFRA (2014a).

Episode	Pollutant(s)	Areas of UK described as being affected
10-14 Jan	PM	Northern Ireland and northwest England
17-25 Jan	PM	Eastern England; Wales and Northern Ireland later
10-28 Feb	PM	Periods of pollution across UK
20-31 Mar	PM	'Widespread'
1-18 Apr	PM & O_3	Most / all of UK
7-8 May	PM & O_3	All of UK
17-20 May	O_3	South of UK
25-31 May	O_3	'Widespread'
19-20 Jun	PM &O_3	South of UK
5-24 Jul	O_3	Most of UK
27 Jul	O_3	Not stated
22-26 Nov	PM	Northern Ireland and central Scotland
10-13 Dec	PM	England and south Wales

CHAPTER FIVE

Current Challenges

Unless the Government is forced to act, air pollution will go on making people sick and causing tens of thousands of early deaths every year in this country.

Alan Andrews, ClientEarth

The sheer scale of air pollution in modern Britain is particularly concerning because of its known associations with distressing health outcomes and significant levels of excess mortality. This serious situation poses major challenges that rapidly require an informed and effective set of responses and this will be the focus of the next chapter. First though, a critical examination of our current approaches to air pollution is likely to be instructive, to identify the main shortcomings that need to be overcome now if real progress is to be made in future.

Is Current Monitoring of Air Pollution Effective?

On the face of it, the UK's national monitoring network, the AURN, is something to be commended. It consists of more than a hundred monitoring stations, all with expensive, state-of-the-art equipment giving accurate readings. Many of the stations monitor for more than one pollutant. They also measure pollutant concentrations continuously, throughout the day and, barring technical downtimes, on every day of the year. The data they record are sent every hour by telemetry to a central database and millions of individual results are stored in perpetuity. Not only that, but these raw data are freely available for anyone to scrutinise. So, in summary, the AURN is comprehensive, scientifically robust and transparent. However, when human health and lives are at stake, it is valid to pose some questions.

Firstly, a national air quality monitoring network is a requirement under EU law: so are successive UK governments simply undertaking the minimum requirements and, if so, is this enough? In fact, the EU Directives on air quality do not specify a required minimum number of monitoring stations across a given area but rather that there should be 'full coverage' for PM_{10}, NO_2 and some other pollutants, since 'all exceedances should be detected' (EEA, 2011). There *is* a specific requirement though for monitoring to take place in agglomerations with more than 250,000 inhabitants. Therefore, under current EU law, a more robust UK monitoring network may not be necessary given the extensive nature of the existing infrastructure and the additional modelling that is undertaken. And, very likely, it may be politically challenging to expand monitoring further since resources are finite and new uses for taxpayers' money inevitably lead to shortages elsewhere unless taxes are to rise (always politically unpopular). Of more concern for the future however is whether the current level of monitoring will be continued in a post-'Brexit' UK that is no longer beholden to EU regulations. Concerned citizens may need to be vigilant, and possibly vociferous, to ensure that environmental protections of these kinds are not watered down.

With regards to current levels of monitoring, a more pertinent question might be: Are the existing monitors all in the right places and fully representative of likely human exposure? For example, in Plymouth, the city where I work, the AURN monitoring site is located in the middle of a large pedestrian precinct some 140 metres away from any vehicular traffic, which realistically is the only potential source of serious air pollution in the city. Meanwhile, some of the roads in the city centre (none of which are monitored by the AURN) are busy thoroughfares for both pedestrians and vehicles. For example, one such road, Royal Parade, is a busy highway with lines of city traffic, separate bus lanes and numerous bus stops with idling, diesel-fuelled buses, all bounded by wide pavements that are always full of shoppers and commuters. If there is likely to be any problem of human exposure to air pollutants in this particular city it will be on Royal Parade and similar locations, not in the pedestrian precinct where any traffic pollution will already have been diluted by the time it arrives there, if it arrives at all.

Therefore, data monitoring sites, like the one in Plymouth, could be giving a false picture of the extent of air pollution citizens are being exposed to. Each AURN monitoring station has a separate web-page on the government's air quality website (www.uk-air.defra.gov.uk) and examination of these indicates that the 140 metre gap in Plymouth between the monitor and the nearest road is fairly typical for the UK's fifty-five urban background monitoring sites[1]; in many cases the distance to the nearest busy road is, in fact, much greater. It should be pointed out here that there are also fifty-four AURN urban *roadside* sites across the country but many towns and cities (not just Plymouth) do not have a roadside monitor, only a background one, and in all these cases the observation made above remains valid. It is also true that local authorities (LAs) also make measurements of air pollution in their areas, including at roadsides, and are required to draw up action plans if concentrations breach the official AQS (see below); however, such LA measurements, unlike those derived from the AURN, are not as intensively collected, do not have the same coverage of air pollutants and, more to the point, are not as readily available to the public.

A final issue to bear in mind with regards to AURN monitoring stations is that the air intake point is typically set at a height of four metres above the ground, as detailed on the individual site descriptions on DEFRA's air quality website. This is generally done for entirely practical reasons, such as keeping the inlet out of the reach of interference by curious or mischievous members of the public. It does mean however that officially declared pollutant concentrations may not always be fully representative of the situation nearer to the ground, where most of the exhaust emissions occur and where pedestrians, small children in particular, are predominantly exposed.

The Trouble with Objectives

We have already seen that pollutant concentrations exceed UK AQSs at nearly all of the UK's air quality monitoring sites (see Table 4.3). Yet, despite this, the UK government is able to declare each year that the EU-set 'objectives' for air quality have been met at most of its monitoring sites. How can this be? The answer lies in the wording of the air quality objectives used throughout the EU, which allow every monitoring site to have a permitted number of AQS 'exceedances' each calendar year; these are listed for the main three pollutants in Table 5.1. For PM_{10} for example, the objective for each monitoring location is that it should have no more than thirty-five exceedances per year. This effectively means that PM_{10} pollution in a UK town can breach WHO guidelines once every ten days or so, year after year, but still be 'meeting its air quality objectives', even

Table 5.1. Permitted numbers of exceedances for air pollutants. The European Union air quality objectives are that specific air pollutants must not exceed a permitted number of exceedances of the limit value (air quality standard) in any one calendar year.

Pollutant	Measurement	Permitted number of exceedances per year
NO_2	Hourly mean	18
O_3	8-hour mean	10[a]
PM_{10}	Daily mean	35

a. For ozone, the objective shown here is that designated by the UK government. The EU objective is for no more than 25 exceedances per year (averaged over three years). However, the EU's long term objective for ozone (non-mandatory at present) is for there to be zero exceedances of the EU target value.

though its residents are regularly being exposed to potentially hazardous levels of air pollution. This cannot be satisfactory when we remember the ill health and thousands of deaths attributed by public health experts to PM pollution each year. Members of the general public, hearing that EU *objectives* have been met for air quality in their area, are likely to take this at face value and believe that there is no problem – that there have been no breaches of the health-driven air quality *standards*; yet in most cases, as seen earlier in Table 4.3, the standards are being routinely breached in the UK.

To illustrate further the potential for objectives to give a misleading picture of the nation's air quality, Table 5.2 lists, for each pollutant, the breaches of both AQSs and objectives in 2014. For the three pollutants shown, the percentages of AURN sites that failed the air quality objectives (final column) are low, giving the impression of relatively good air quality in the UK. However, most AURN sites

Table 5.2. Comparison between exceedances of air quality standards and objectives in UK, 2014[a]. Annually measured NO_2 and PM_{10} not included in the table as the objective in these cases is simply to meet the annual AQS.

Pollutant	Measure-ment	Monitoring stations exceeding UK AQS	Monitoring stations failing objective[b]
NO_2	Hourly mean	14% (17/124)	<2% (2/124)
O_3	8-hour mean	89% (71/80)	18% (14/80)
PM_{10}[c]	Daily mean	84% (55/65)	0% (0/65)

a. Shown in brackets are the numbers of monitoring stations exceeding the AQS on one or more occasions (third column) and those failing the air quality objective (fourth column) out of the total number of UK stations that were monitoring each pollutant in 2014.

b. Objective: for AQS not to be exceeded on more than a permitted number of occasions per year (see Table 5.1).

c. For $PM_{2.5}$ there are no statistics on exceedances because a UK AQS had not been set for this pollutant by 2014.

across the UK in fact exceeded the health-focussed AQSs (third column). This is most starkly shown for PM_{10} which did not fail the objective at a single UK monitoring site, despite the vast majority (84%) of the same sites having one or more exceedances of the AQS during the year. For ozone, 89% of the monitoring sites (including at rural locations) exceeded air quality standards yet only 18% failed the air quality objective. Similarly, for NO_2, the government can state that the UK met its EU objectives for hourly means at all but two of its monitoring sites, yet seventeen stations recorded breaches of the AQS on one or more occasions. The *annual* mean for NO_2 (not shown) was breached at a similar number of monitoring locations (nineteen), further indicating a more widespread problem for NO_2 than the figures on air quality objectives suggest. Furthermore, the official figures on UK objectives are based only on monitoring sites within the AURN, whereas separate monitoring activities show that more locations would be in breach if they were included. For example, some sites within the London Air Quality Network (ERGKC, 2016a) have failed the NO_2 objective (no more than eighteen exceedances of the hourly AQS in a calendar year) after only a few days of the new year. In 2016, the objective had already been failed at Putney High Street in London by January 8th and in 2015, the objective had been failed at Oxford Street after only two days of the year (ERGKC, 2016c). By contrast, in neither of these time periods were there any exceedances of the AQS at London Marylebone Road, the AURN monitoring site with some of the worst NO_2 problems.

Are Air Quality Standards Protective of Health?

The focus so far in this chapter has been on breaches and exceedances of the air quality standards and it might seem fair to assume that residents living in areas that are never in breach may be safe from pollution-related health problems. Unfortunately, a closer look at the science behind the setting of AQSs shows that they can involve a large element of practicality and that meeting them is by no means a guarantee of public health safety. As an example of this, let's consider the WHO's guideline for PM_{10} of 20 μg m^{-3} as a yearly average (Table 5.3). With this guideline (and the one for $PM_{2.5}$) the WHO aims to protect against excess mortality risk from PM exposure, particularly mortality associated with cardiopulmonary disease and lung cancer (WHO, 2006). It may come as a surprise then that the UK's AQS for PM_{10} as an annual average is 40 μg m^{-3}, double the WHO recommendation. This means that levels of PM_{10} in the UK atmosphere can be considerably higher than the WHO safety guideline yet still be described as meeting national air quality standards. In contrast, the table shows that for NO_2 and O_3, the UK AQSs are set at the same concentrations as the WHO recommended guideline levels.

Table 5.3. Guidelines and air quality standards for the main air pollutants of current public health concern (all in units of μg m^{-3}). The UK AQS are broadly those of the EU except for O_3, which has an EU target value of 120 μg m^{-3}. Figures in bold indicate where the UK air quality standards (AQS) are set at levels that do not meet the WHO air quality guidelines (AQG),

Pollutant	Measurement	WHO-AQG	UK AQS
NO_2	Hourly mean	200	200
NO_2	Annual mean	40	40
O_3	8-hour mean	100	100
PM_{10}	Daily mean	50	50
PM_{10}	Annual mean	**20**	**40**
$PM_{2.5}$	Daily mean	**25**	**None set**
$PM_{2.5}$	Annual mean	**10**	**25**

In setting its air quality guidelines, the WHO appears to acknowledge that some pollution, and the ill-health and mortality it causes, is inevitable at present[2]. Certainly, given our current lifestyles, and our dependence on the internal combustion engine in particular, reducing the concentrations of pollutants like PM_{10} to safer levels will require large political and cultural shifts. Accordingly, the WHO has set a series of interim targets for PM_{10} (and $PM_{2.5}$) to be met by countries like the UK that have not yet committed to the recommended 20 μg m^{-3} level[3] (Table 5.4). The WHO stresses that exposure to PM_{10} at the higher, interim target concentrations carries additional mortality risks. Their explanations of the interim targets (final column of the table) indicate that the weaker UK AQS of 40 μg m^{-3} carries additional mortality risks of between 3% and 9%[4], compared to the WHO-recommended level of 20 μg m^{-3}. In other words, wherever annually-averaged PM_{10} concentrations are between 20 and 40 μg m^{-3}, UK citizens are being exposed to these additional mortality risks despite official advice that pollutant levels in their neighbourhood are meeting the government's official air quality standards.

For $PM_{2.5}$ as well, the newly adopted EU and UK AQS of 25 μg m^{-3} (bottom row, Table 5.4) is noticeably laxer than the 10 μg m^{-3} that has been recommended by the WHO. The WHO recommendation is again based on evidence that increased health effects and mortality occur at levels above this. As for PM_{10}, the WHO has recommended interim guidelines for annually-averaged $PM_{2.5}$ that carry additional mortality risk. At a concentration of 25 μg m^{-3} (the UK AQS) the additional mortality risk is 9%. In the meantime, the UK has, at the time of writing, no AQS at all for short-term (daily) $PM_{2.5}$ exposure, despite the WHO-recommending a separate guideline level for this of 25 μg m^{-3} (not shown in the table) based on short-term health and mortality risk.

Is the Public Adequately Informed?

Setting aside for a moment the limitations of the EU and UK air quality standards and objectives, it is arguable that any breaches of either ought to be communicated to the public; however, the official EU requirement is only that the public should be informed if higher, 'alert' thresholds are exceeded[5]. As it is, breaches of air quality standards and objectives are not widely publicised and the average British citizen is therefore unlikely to be aware when they occur – and we have seen that they occur routinely every year (e.g. Table 4.3). One particular air pollution episode in 2014 was covered by TV

Table 5.4. WHO interim targets and air quality guidelines for particulate matter, compared to the UK air quality standards. All figures are expressed as annual averages. Source (except bottom row): WHO, 2006.

Targets and standards	PM_{10} ($\mu g\ m^{-3}$)	$PM_{2.5}$ ($\mu g\ m^{-3}$)	Basis for the selected level (WHO text)
Interim target 1 (IT-1)	70	35	These levels are associated with about a 15% higher long-term mortality risk relative to the AQG level.
Interim target 2 (IT-2)	50	25	In addition to other health benefits, these levels lower the risk of premature mortality by approximately 6% (2-11%) relative to the IT-1 level.
Interim target 3 (IT-3)	30	15	In addition to other health benefits, these levels reduce the mortality risk by approximately 6% (2-11%) relative to the IT-2 level.
Air quality guideline (AQG)	**20**	**10**	These are the lowest levels at which total, cardiopulmonary and lung cancer mortality have been shown to increase with more than 95% confidence in response to long-term exposure to $PM_{2.5}$.
UK AQS for comparison	40	25	

news and national newspapers but, most likely, only because it coincided with a plume of Saharan dust that made the pollution *visible*. In most cases, there is a sense that interested parties have to go looking for air quality information, on government websites perhaps, and that everyone else will remain blissfully unaware of any major pollution, despite the harm it may be doing to their health.

The WHO guideline levels and UK air quality standards discussed in detail in this book are not commonly known to members of the British public, who need a simple system to tell them whether or not their health is at risk from air pollution. For the UK, DEFRA, with advice from the independent Committee on the Medical Effects of Air Pollutants (COMEAP), has established a daily air quality index (Table 5.5) that divides air quality into four bands: low; moderate; high and very high, based on a wider 1-10 scale, where 1 equates to the lowest pollutant levels and 10 to the highest. The index is used by DEFRA in air quality forecasts and in retrospective statistics on past pollution. The four bands carry official advice (Table 5.6) so that, in theory, concerned members of the public can make informed decisions about their activities in relation to the prevailing or forecast air quality. COMEAP has also recommended separate 'trigger concentrations' for short-term spikes in PM and O_3 levels, to better inform air quality forecasting.

The air quality index, and the expert advice that has produced it, are to be commended. But the concern raised above – that air quality information is not particularly well *publicised* – applies here too; whilst interested individuals can access this excellently presented information online, if they know where to look, its existence is not widely known. For example, the air quality forecasts based on the index are generally not included in national weather forecasts on television. This would surely not be the case if there was sufficient awareness of air pollution and its dangers; after all, information on pollen and UV levels appear routinely on television during summer[6]. Excellent efforts have been made recently to provide individuals with the option of receiving air quality updates via social media[7]; this is a very promising method of widening awareness but does not appear to be widely promoted and will therefore be of use only to those who are aware that such initiatives exist.

Table 5.5. The UK's air quality index. All concentrations in units of μg m⁻³.

		Ozone	Nitrogen dioxide	Sulfur dioxide	$PM_{2.5}$	PM_{10}
		Averaging period				
Band	**Index**	**8 hrs**	**1 hr**	**15 mins**	**24 hrs**	**24 hrs**
Low	1	0-33	0-67	0-88	0-11	0-16
	2	34-66	68-134	89-177	12-23	17-33
	3	67-100	135-200	178-266	24-35	34-50
Moderate	4	101-120	201-267	267-354	36-41	51-58
	5	121-140	268-334	355-443	42-47	59-66
	6	141-160	335-400	444-532	48-53	67-75
High	7	161-187	401-467	533-710	54-58	76-83
	8	188-213	468-534	711-887	59-64	84-91
	9	214-240	535-600	888-1064	65-70	92-100
V.high	10	≥ 241	≥ 601	≥ 1065	≥ 71	≥ 101
AQSs[a] for comparison[b]		100	200	266	25	50

a. AQS = Air quality standards.

b. Concentrations shown in this row (not part of the air quality index but added here for comparative purposes) are the UK air quality standards, with the exception of $PM_{2.5}$, for which there is no UK-AQS for *daily*-measured concentrations; the WHO air quality guideline for daily $PM_{2.5}$ is shown here instead.

Table 5.6. Official advice relating to air quality index (Table 5.5).

Band	Index value	Accompanying health messages for at risk individuals	Accompanying messages for the general population
Low	1-3	Enjoy your usual outdoor activities.	Enjoy your usual outdoor activities.
Mod-erate	4-6	Adults and children with lung problems, and adults with heart problems, who experience symptoms, should consider reducing strenuous physical activity, particularly outdoors.	Enjoy your usual outdoor activities.
High	7-9	Adults and children with lung problems, and adults with heart problems, should reduce strenuous physical exertion, particularly outdoors, and particularly if they experience symptoms. People with asthma may find they need to use their reliever inhaler more often. Older people should also reduce physical exertion.	Anyone experiencing discomfort such as sore eyes, cough or sore throat should consider reducing activity, particularly outdoors.
Very High	10	Adults and children with lung problems, adults with heart problems, and older people, should avoid strenuous physical activity. People with asthma may find they need to use their reliever inhaler more often.	Reduce physical exertion, particularly outdoors, especially if you experience symptoms such as cough or sore throat.

Precisely because the air quality index is not widely publicised, or therefore known by the general public, there is a danger that the information it contains will be misconstrued. For example, in air quality forecasts, reference is made to one of the four bands so that a particular day is forecast to have a low, moderate, high or very high level of air pollution. But a pollutant concentration that *exceeds* the air quality standard (shown in the bottom row of Table 5.5) results in a classification of 'moderate' air pollution or even, in the case of $PM_{2.5}$, 'low'. It is a distinct possibility that most people, unaware of the health messages that are linked to these classifications (Table 5.6), might think there is no pollution problem unless they see a 'high' or 'very high' classification. The majority of people may well not realise that a classification of 'moderate' for the air quality in their area actually means that concentrations are higher than those recommended for short-term exposure by the World Health Organisation.

The foregoing is not to criticise the air quality index but, in fact, the failure to promote its widespread use, considering the routine breaches of air quality standards that occur in Britain. However, there is some basis for questioning the classification system for $PM_{2.5}$. For this pollutant, a forecast concentration within the range of 26-35 $\mu g\ m^{-3}$ is classed in the index as being in the 'low' air pollution band, despite it being in breach of the WHO recommended guideline for daily-measured $PM_{2.5}$ of 25 $\mu g\ m^{-3}$. This seems remiss and should perhaps be reviewed, particularly when it is remembered that $PM_{2.5}$ is known to be particularly damaging to health and forms the basis of the oft-quoted figure of 29,000 attributable deaths per year in the UK.

In summary, whilst the daily air quality index is generally an excellent development – and based on the recommendations of public health experts – it should be more widely broadcast, together with accompanying health advisories. Until this happens, there is the unfortunate possibility that the index will be of little help to members of the general public, who will not all have the time or inclination to do research on the details of the index and how the various bands relate to the WHO guidelines, UK AQSs and health advisories.

A Little Local Difficulty

The UK has a national air quality strategy, which the EU requires all member states to produce. Its main focus is on human health impacts and it details the national air quality standards and objectives and the air pollutants of most concern. One of key ways in which the national strategy is meant to be delivered is local air quality management (LAQM). Under the system of LAQM, every one of the 389 local authorities (LAs) in the UK has a duty to monitor air pollution in its area. If the LA's monitoring indicates that air quality objectives are not being complied with in a specific location, or are not likely to be, the LA has to declare an air quality management area (AQMA) for that location. Each AQMA has to cover the area affected by pollution; this may be an isolated location, such as a particularly congested street, or a wider area covering a busy city centre perhaps. Once an AQMA has been declared, the LA has to write and deliver an air quality action plan within twelve months of the declaration. The action plan details the nature of the pollution problem, including the identified sources and causes, and includes suggestions of measures that should be taken to alleviate the identified problems. By 2014, the majority of UK LAs (257) had declared at least one AQMA and nationally there were 685 AQMAs; the vast majority of these (581) were for NO_2, probably because NO_2 monitoring is carried out routinely by LAs, using relatively simple and cheap 'diffusion tubes', whereas measurements of other pollutants, particularly fine PM, require more expensive equipment.

Local air quality management is an admirable system, with air quality officers across the country working hard to figure out the best ways to reduce pollution in their local areas. However, the actions they identify generally require resources and support from central government. Traffic congestion is typically the source of the worst problems but LAs do not have the means to reduce the sheer amount of traffic on Britain's roads; it requires broader solutions that can only come from central government. Where specific solutions are identified locally, to ease congestion on particularly affected roads for example, greater financial support is typically required to achieve the desired outcomes but this has looked unlikely to be forthcoming as public spending has increasingly been cut to reduce the national deficit. In fact, the UK government has previously proposed plans to abolish LAQM in a bid to cut administration and costs; on that occasion it relented in the face of opposition but financial pressures have not disappeared. Clearly there are always going to be heavy demands on taxpayers' money but it is worth remembering that air pollution has large financial costs that would be dramatically reduced if its root causes were adequately addressed. The European Commission estimates that the direct costs of air pollution to the EU are €23 billion per year but the indirect costs of health outcomes amount annually to €330-940 billion (EC, 2013). In the UK, the costs of air pollution in London alone could be up to £3.7 bn per year (Walton et al., 2015).

The 'Dash for Diesel' and 'Dieselgate'

Much of Britain's current pollution stems from the large number of vehicles on the nation's roads. Apart from electric and hydrogen fuel-cell cars[8], all vehicles emit pollution at source, leaving both motorists and pedestrians exposed to the fumes. Particularly problematic in pollution terms are diesel-powered vehicles. Diesels emit more NO_2 than petrol vehicles[9] because only the latter are fitted with a three-way catalytic converter that removes 99% of NO_2 from exhaust gases. Three-way catalytic converters have been standard on petrol vehicles for many years but are not appropriate for diesel vehicles because excess oxygen in the exhaust, resulting from the diesel's lean-burn engine, would disrupt the reduction of NO_2. As well as their higher NO_2 emissions, diesel engines produce significantly more fine and ultra-fine PM than their petrol counterparts[10] because they are prone to incomplete fuel combustion and this results in the emission of black carbon particles. In addition to the health and mortality impacts of the NO_2 and PM emitted by diesel vehicles (see chapter 3), diesel exhaust *in general* has been classified by the International Agency for Research on Cancer as carcinogenic, with sufficient evidence that exposure to diesel fumes can cause lung cancer and, possibly, bladder cancer (IARC, 2014).

It is therefore highly problematic that the number of diesel vehicles on British roads has risen dramatically in recent decades, from less than 2 million in the 1990s to nearly 11 million by 2015; they now account for more than half of new cars and one-third of all road vehicles (DfT, 2015a). The main reason for the rapid increase is that diesels have been a financially attractive option, partly because of their fuel efficiency; they are 20-30% more efficient than similar petrol vehicles, significantly reducing the cost of filling up with fuel. Also, their fuel efficiency means they emit less carbon dioxide, the main greenhouse gas implicated in global warming and, for this reason, the Labour government in 2001, aiming to meet its carbon-reduction commitments, introduced a new road tax regime that encouraged the purchase of the most energy efficient vehicles; this taxation system was applied by successive governments and the number of diesel vehicles on our roads grew significantly. Encouraging the wider use of diesel was a relatively easy way of addressing a specific environmental problem whilst allowing the population to carry on with 'business-as-usual' and even save money in the process. Ultimately though, the only way to really address carbon emissions (short of adopting untested and possibly risky carbon capture schemes) will be to stop burning fuels of any kind, including diesel. And here, perhaps, lies a cautionary tale that we should reflect on when considering how to solve our air quality problems; quick or easy fixes are unlikely to really help in the long run and more difficult and unpopular choices may ultimately be necessary (see Chapter 6).

The polluting nature of diesels came to global attention in 2015 with news that Volkswagen, the world's largest car manufacturer, had installed a so-called 'defeat device' into the engine software of 11 million diesel cars. This was designed to significantly reduce NO_2 outputs from the exhaust – but only when the software detected that the vehicle was undergoing a standard emissions test. Under normal driving conditions, the NO_2 reduction technology (based on removing NO_2 by reacting it with added ammonia or urea) was switched off to improve the vehicles' fuel economy figures. Emissions from these vehicles were found to be up to forty times the permitted levels, despite passing official emissions tests.

The Volkswagen 'dieselgate' case drew global attention to the air quality problems associated with diesel engines and also to broader questions about the veracity of emissions tests in general, not just those relating to Volkswagen. Vehicle emissions tests are held under laboratory conditions and had actually been known for some time to be largely unrepresentative of 'real-world' emissions; this has also been borne out by subsequent testing, which has shown other manufacturers' diesel vehicles to be routinely failing emissions standards in real-world conditions, despite officially being compliant. National governments, including the UK's, have been criticised for 'turning a blind eye' to this state of affairs and not taking action against big business in order to protect public health. Perhaps the approach would be more robust if air quality was a higher priority in the public eye and the political arena.

Government Challenges

In 2008, the European Commission's Air Quality Directive committed EU member states to meeting EU limit values for air pollutants by 2010. The UK, along with a number of other EU states, did not meet this commitment, with NO_2 pollution being particularly problematic. For several years now, the UK government has been threatened with fines by the EU for its continued failure to comply, although no fine has ever been levied. In 2011, a House of Commons committee published an air quality report that was critical of the government, which it said was trying to avoid EU fines by behind-the-scenes lobbying to 'dilute' EU air quality standards (House of Commons Environmental Audit Select Committee, 2011); the chair of the Committee described the situation as "a national scandal". A clause in the 2008 Directive contained a provision for possible time extensions of up to five years for compliance and the government's response to failures on NO_2 pollution has been to request much longer time extensions for 38 of its 43 air quality zones. Its current projections (DEFRA, 2014b), indicate that compliance will not be achieved until 2020 (in ten of the zones), 2025 (in a further twenty-three zones), 2030 (two zones) or *after* 2030 (three zones: Greater London, West Midlands and West Yorkshire).

In response to the ongoing and projected failures in compliance, ClientEarth, a non-profit organisation of lawyers, launched a legal challenge against the UK government, claiming that it had not "considered or put in place all practical measures to ensure compliance" (Supreme Court, 2015). The UK Supreme Court ruled in ClientEarth's favour and ordered the government to submit a new National Plan to the European Commission. When it subsequently launched the new plan, the government stated that in response to the Supreme Court ruling, there would be 'Clean Air Zones' (CAZs) in five UK cities (Birmingham, Derby, Leeds, Nottingham and Southampton) (DEFRA, 2015c). In each of the CAZs, diesel taxis, buses, coaches and lorries would be discouraged from entering the city centre by the levy of charges, to be introduced in 2020. On every level, this can only be described as a half-hearted response. There was no action on private cars, no action in many of the locations that are also in breach of pollution guidelines (e.g. Aberdeen, Bath, Belfast, Caerphilly, Glasgow and Oxford), no additional actions for grossly-polluted London, no action for five years following the ruling and, furthermore, the polluting vehicles targeted were not, in fact, to be prohibited from affected areas, only 'discouraged' via entry charges. Despite the Supreme Court ruling, and its vindication of ClientEarth's original action, there still appears to be no strict timetable for final compliance on NO_2 limit values nationwide, or indeed any sanctions for the existing and ongoing failure to meet the legal obligations that were set originally for 2010.

A separate challenge for the UK government lies in its responses to ongoing emissions of greenhouse gases. Whilst the focus in that case is separate from local air quality, any proposals to reduce our dependence on fossil fuels are clearly of relevance in tackling not only CO_2 and other greenhouse gases but also the most problematic local air pollutants such as NO_2 and PM. In 2015, following many years of unsatisfactory international meetings on global warming, world leaders finally agreed at talks in Paris to shift away from fossil fuels by mid-century. Despite the rhetoric of UK representatives at such meetings, there remains scepticism about the government's level of commitment to decarbonisation and the improved air quality it would undoubtedly bring. Contradictory policies announced by the government in the same year included the following examples: reduction of subsidies for solar panels; withdrawal of rules for new homes to be carbon neutral; removal of financial incentives for the purchase of less polluting vehicles; and the granting of licenses for shale gas exploration.

In summary, the UK government is not only in breach of legal air pollution limits, but openly expects to remain so for some considerable time. This raises grave concerns when we consider the serious and very real health risks associated with emissions from vehicles and other sources. Achieving good air quality is a major challenge that will require far deeper commitments in the coming years and there will need to be radical changes ahead if the deadly air pollution in our towns and countryside is to be eradicated. As described in the following chapter, this will require a level of action and leadership from our elected politicians that is so sadly-lacking at the moment, together with the adoption of novel technologies and, perhaps most important of all, fundamental adaptations to our current ways of living and working.

CHAPTER SIX

Clearing the Air

You must be the change you wish to see in the world
Attributed to Mahatma Gandhi

The Challenge Ahead

Our way of life in the UK, particularly the dependence many of us have on private vehicles, means that air pollution is still an everyday occurrence in post-industrial Britain. But continuing with business-as-usual does not appear to be a sensible option when we consider the serious health implications of our poor air quality. Cleaning up our air will require current and future governments to take far more effective and sustained actions than we have seen so far. As it is, there is a feeling amongst those concerned about air pollution that it is not given sufficient attention by our political leaders. For example, the respected national charity, Environmental Protection UK, points out that the country's National Air Quality Strategy (an EU requirement) has not been updated since 2007 and "lacks ambitious targets and actions" (EPUK, 2015).

Importantly, any central government policies and actions on air quality will first require general acceptance and support from the people – the electorate – so that politicians feel emboldened to act in the common interest without risking a slide in the opinion polls or, worse still from their perspective, defeat at the next general election. With this in mind, it is almost certain that methods and approaches for tackling air pollution will be adopted only if they do not limit personal freedoms and do not detract from the comfort and convenience that UK citizens currently enjoy. Poor air quality could be described as an 'inconvenient truth' – to borrow Al Gore's phrase about global warming – and, as such, is not likely to be solved by electorally-unpopular sledgehammer approaches. An important added element here though is the apparently low level of public awareness about the serious health impacts of our largely invisible air pollution. If the average UK citizen had a full understanding of the severity of the problem, then more radical actions to tackle air pollution would probably have a greater chance of success.

As with personal freedoms, economic realities cannot be cast aside in our quest for a clean air utopia. As a general rule, and more than ever since the global financial crash of 2008, proposals for any perceived beneficial cause have to be tightly costed and must be affordable from a shrinking public purse. The good news here is that achieving a country without gross air pollution would have significant financial benefits, not least because of the billions of pounds that would be saved every year by reducing the ill health and deaths associated with smoggy air. A study by researchers at the Massachusetts Institute of Technology estimates the 'total monetized life loss in the UK' due to air pollution to be between £6 billion and £62 billion per annum, equating to up to 3.5% of gross domestic product (Yim and Barrett, 2012). Estimates of the global costs of mortality impacts are in excess of 3.5 trillion US dollars in lost labour and increased welfare costs (World Bank Group and IHME, 2016). The challenge here however is incorporating far-sighted estimates such as these into conventional economics.

An important wider point is that action in the UK alone, however comprehensive it might one day be, can never be the whole solution to our air pollution problems. This is simply because much of the ill-health and loss of life expectancy in the UK is attributed to pollution imports from abroad, mainly continental Europe. Researchers have estimated that almost a third of the annual UK deaths attributed to air pollution are due to continental pollutants being blown across the English Channel (Yim and Barrett, 2012). To this end, pan-European initiatives, such as the 2013 EU Clean Air Programme for Europe (which contains proposals for new air quality objectives, Directives and urban improvements) are vital and should be vigorously pursued, regardless of the UK's political position in Europe; as should broader international programmes such as the 1979 Convention on Long-Range Transboundary Air Pollution (CLRTAP), under which fifty-one states formally recognised the need to seek solutions and cooperate on developing appropriate policies, research programmes and monitoring techniques. International action on greenhouse gas emissions is also relevant because this will mainly require reductions in the combustion of fossil fuels, the main source of poor air quality, as well as anthropogenic global warming. Again though, meaningful international actions are only likely to be implemented with the full support of the voters of individual countries.

With these important points in mind, it is clear that any initiatives and actions to clean up the UK's air pollution will have the greatest chance of success if they satisfy a number of conditions. They should be practicable. They should be cost-neutral, once all cost savings (mainly from improved public health) have been taken into account. They should carry the support of a government that is serious about protecting its citizens from unavoidable, indiscriminate and potentially serious damage to their health. And they should be accompanied, or preceded, by a full and frank programme of public information and education about the serious health problems associated with poor air quality, so that any proposed actions are more likely to be accepted, and even demanded, by the voting population. Finally, appropriate actions to improve air quality should, ideally, not be taken in isolation in individual countries like the UK but as part of coordinated international programmes that recognise the reality of transboundary air pollution.

Approaches to Poor Air Quality

In practical terms there are two key approaches that are likely to be most successful in achieving meaningful reductions in air pollution, and both require central government support and coordination, underpinned by national policies. Firstly, the widespread adoption of proven and affordable clean technologies that are already available to us. The second imperative is for a *fully-informed* public to engage in a perceptible cultural shift in attitudes and behaviours, facilitating a move to more sustainable lifestyles that acknowledge the environmental and financial costs of business-as-usual and the fundamental benefits of a cleaner and healthier environment; only this will ensure a sufficiently enthusiastic adoption of clean technologies and the necessary changes to some of our current ways of doing things. With these thoughts in mind, let's consider how the various sources of air pollution might be addressed.

Reducing Pollution from Road Transport

In post-industrial countries like Britain, the most harmful episodes of air pollution, particularly in urban areas, tend to be caused by transport emissions. There are two broad approaches that can be taken to fundamentally address the problem, one based on cleaner vehicles and one based on fewer vehicles. These approaches are by no means mutually exclusive and action on both fronts would be beneficial; however, moving towards cleaner, rather than fewer, vehicles is a more likely option unless British citizens willingly give up the private car travel that so many enjoy or aspire to. Moving to fewer vehicles is much more radical and would rely on fundamentally discouraging and reducing the mass use of private vehicles, partly by providing reliable and efficient public transport. Whilst discouraging the use of private vehicles would undoubtedly be unpopular it would actually be more effective at reducing *total* pollution emissions and environmental damage, when the environmental impacts of mass vehicle manufacture are also taken into account. Our current culture, where large, resource-intensive vehicles often carry just one or two occupants (sometimes on short and perhaps walkable journeys), could hardly be described as environmentally sustainable.

Cleaner Vehicles

Let's first consider the opportunities for clearing the air and improving our health without ending the UK's current dependence on the private car. Emissions from vehicles with an internal combustion engine can be reduced in two ways; by utilising the best, widely-available emissions control technology and by encouraging the purchase and use of the least-polluting models currently available, using financial incentives. In the longer-term, a more beneficial approach would be the more widespread adoption and use of cars (and other vehicles) using electricity or fuel cells, which do not emit any pollution from the exhaust. We will consider these approaches one at a time, starting with technology based on the internal combustion engine.

Petrol vehicles emit exhaust gases and fine particles, particularly under certain driving conditions; however, progress has already been made, particularly for NO_2, one of the most harmful urban air pollutants. Since 1993, every new petrol car in the EU has been fitted with a three-way **catalytic converter**, which significantly reduces emissions of NO_2 as well as other pollutants (carbon monoxide and unburnt hydrocarbons) and, in this context, petrol vehicles are certainly cleaner than diesels. That said, petrol vehicles are, in no sense, 'zero or low emission'; after all, they rely on internal combustion of fossil fuel and generate waste gases and particles that are emitted into the atmosphere adding significantly to air pollution.

For reasons described earlier, diesel engines emit higher levels of both NO_2 and fine PM, compared to petrol vehicles. For NO_2, this is mainly explained by the fundamental incompatibility of diesel engines with the catalytic converters that minimise NO_2 emissions from petrol vehicles. Whilst alternative technological methods can be employed to reduce diesel NO_2 emissions, their use can reduce vehicle performance and this is what prompted the Volkswagen scandal of 2015 (see Chapter 5). In the case of fine PM, emissions can be reduced by **diesel particulate filters** (DPFs), which are fitted to modern diesel vehicles in the UK to meet EU emission limits. Cars have been fitted with DPFs since 2011 and, effectively, they have been required in buses and lorries since 2006, also to meet emission limits. Unfortunately, there are inherent problems associated with the use of DPFs. The filter is regenerated when captured soot particles are burnt off the filter at high engine temperatures; however, on repeated short journeys (e.g. to and from a nearby workplace) the engine may never warm sufficiently and the DPF can become blocked, requiring repair or replacement at a typical cost of over £1000. This has led to many car-owners having their vehicle's DPF illegally removed. Whilst that saves money for some motorists, and gives business to DPF removal companies (many of which can be seen advertising online), it leads to higher PM emissions to the atmosphere.

Another technological innovation already in use in transportation is the **hybrid vehicle**, where a traditional internal combustion engine provides power, not only to the wheels, but also to a battery that stores excess energy; the stored power can then be used when less motive power is required, at lower speeds for example. Because less fuel is burnt, exhaust emissions are lower than a conventional fuel-powered vehicle. Despite hybrid cars being relatively expensive, mainly because of the cost of the high-capacity batteries required, they have already become fairly widespread; more than 40,000 were sold in the UK in 2015 compared to >2.5 million conventional cars (SMMT, 2015).

An alternative, non-technological approach to reducing emissions from current, internal combustion vehicles is to offer **financial incentives** for those motorists driving the cleanest, currently-available models. The last few UK governments, with greenhouse gas emissions targets in mind, have set lower road taxes for models (both petrol and diesel powered) with the lowest carbon emissions, providing a financial incentive to motorists to choose a vehicle that emits relatively low levels of air pollutants. Unfortunately, in 2015, the government announced that this incentive was to be removed, indicative of political thinking that does not appear to value environmental and human health protection as particular priorities.

Another alternative approach that does not rely on technology is to **reduce speed limits** on busy roads, which can reduce emissions of key air pollutants because less fuel is burned per mile. However, such measures are not universally popular amongst motorists, who may not fully understand the health benefits of reducing air pollution, or care enough about them when weighed against the perceived or real costs of longer journey times; although it should be noted in this context that temporarily reducing speed limits on motorways is often used at busy times as a tool to ease congestion and *shorten* journey times. Competing interests of this sort can mitigate against adoption of air pollution management and this was illustrated by the ultimate cancellation of plans to reduce speeds on a stretch of the M1 motorway, which had been proposed to improve local air quality problems. Unfortunately, some other approaches that *have* been adopted to reduce vehicle speeds, for valid safety reasons, may be counter-productive in the context of air quality; for example, deceleration and acceleration over 'speed bumps' can increase emissions, leading to poorer air quality, usually in residential streets.

One of the solutions that is sometimes proposed to reduce greenhouse gas emissions from internal combustion engines is the use of '**cleaner fuels**'. These include biofuels, derived from plant materials and wastes, liquefied petroleum gas and compressed natural gas. The advantages of using such fuels, particularly biofuels, is not always clear cut however and, unfortunately, their use in passenger vehicles does not have the effect of significantly improving local air quality (compared to petrol engines) because problematic air pollutants, particularly NO_2 and hydrocarbons, are still emitted.

Despite the availability and adoption of the pollution control technologies and approaches described above, the traditional motor vehicle is still responsible for much of the UK's air pollution and associated ill health. If we are to address the problem of emissions from transport, without phasing out private vehicle use, vehicles will be required that do not depend on any internal combustion of petrol or diesel.

The first widely-available alternative to the internal combustion engine is the **electric vehicle**, which has a large-capacity battery powered by ordinary mains electricity. Private passenger vehicles can be plugged in to a home charging point or one of the 7,000 public charging points now available around the UK. A small proportion of the public charging points (about 500) are fast chargers, requiring about twenty minutes for a 50% charge. Using electricity eliminates the need for any on-board fuel combustion and associated exhaust emissions[1]. If an electric vehicle uses electricity that has been generated by fossil fuel combustion in a power station, then the same pollutants of concern are still emitted into the atmosphere of course, just in a different location, although there are some mitigating factors: for example, some air pollutants can be more efficiently controlled in a power station (using large-scale technologies) than in a small vehicle; secondly, pollutants are likely to be emitted from a tall stack, well away from ground level, allowing for pollutant dispersal and dilution over a large area; thirdly, power stations are generally built away from main population centres, reducing (although by no means eliminating) the risks of widespread human exposure. Despite these mitigating factors however, power stations fired by fossil fuels still emit air pollutants into the atmosphere in large quantities. To avoid the situation where electric vehicles simply transfer air pollution from one place to another, they will need to be powered by renewable or nuclear energy, power sources that will be discussed in the next section.

Generally, take-up of electric vehicles is relatively limited at present, with annual UK sales of fully electric cars in 2015 of less than 10,000 (SMMT, 2015) – much less than 1% of all car sales. The main deterrents to prospective purchasers of electric cars appear to be limited mileage range and high purchase prices compared to conventional vehicles. The limited mileage available from a full charge causes 'range anxiety' in some owners and prospective purchasers, although the network of charging points has grown and will no doubt continue to do so as the electric car becomes more mainstream. Additionally, battery technology is continually improving and this is likely to make electric cars a more attractive proposition to motorists. A further deterrent has been the relatively high purchase cost but prices have decreased over the years and financial incentives are available via government grants of up to £4,500 towards the purchase of a new electric car. The relatively high costs of electric vehicles are, in large part, associated with their batteries; however, battery costs have already decreased by approximately 8% per year since 2007 (Nykvist and Nilsson, 2015) and costs are likely to continue falling. The specialist manufacturer of electric vehicles, Tesla, announced in 2015 that its latest model would be produced in significantly greater numbers than its predecessors and at a more affordable price; hundreds of thousands of advance orders were placed.

The other main alternative to the internal combustion engine in vehicles is the **hydrogen fuel cell**. The fuel cell is the equivalent of the conventional battery found in electric vehicles; both depend on a chemical reaction to provide electricity to power to the vehicle. The main difference is that the reactants in a fuel cell, hydrogen and oxygen, are supplied from external sources (a hydrogen tank and incoming air, respectively) rather than being stored inside the cell. The only chemical product of the reaction is pure water, which is emitted harmlessly as vapour from the exhaust, with no pollutant emissions. Furthermore, hydrogen is much more energy efficient than petrol or diesel; up to 60% of the fuel's energy is converted into motive power compared to less than 20% efficiency in a conventionally powered vehicle (USDE, 2006). The hydrogen fuel cell has existed for many years but has always been hailed as a technology of the future. The main obstacles to its practical use in vehicles have been the low density of hydrogen, necessitating a large and heavy vehicle, and the need for catalysts based on rare and expensive materials such as platinum[2]. However, car makers seem to be finally approaching the point where they can make hydrogen-powered cars commercially viable and, in 2014, the Toyota Mirai became the world's first mass-produced hydrogen car. It is rather heavy, at nearly two tonnes, and very expensive, at a UK launch price of £66,000, but it has a much better range than most electric cars (>300 miles) and no charging time – it just needs to be refilled with hydrogen fuel. Other car makers, including Hyundai, Honda and Nissan are also entering the market or are on the verge of doing so and, in time, prices are likely to come down to more affordable levels, as has happened with electric cars since they were first launched.

For hydrogen vehicles the key issue that needs to be addressed, in the context of environmental protection, is the source of the hydrogen fuel. Currently, the vast majority is produced from natural gas (methane, CH_4); this requires the extraction of a fossil fuel, the input of energy to extract hydrogen from the methane and the process produces CO_2, a greenhouse gas. However, hydrogen can be produced much more sustainably by electrolysis, requiring only liquid water and potentially producing no air pollutants; in this process electricity is used to split water into its constituent hydrogen and oxygen. As with electric vehicles, the key is to use a power source that does not rely on the combustion of fossil fuels (see next section). This is doubly important here because energy is also required for compression of the low-density hydrogen fuel. The widespread manufacture and use of hydrogen cars may only just be starting but there are already signs that the infrastructure required will follow, as we are already seeing for electric cars. The UK government has plans for a network of fifteen hydrogen filling stations in southeast England and, in 2015, the first zero emissions hydrogen filling station in the UK opened in Yorkshire using a dedicated wind turbine to power hydrogen production from the electrolysis of water. It can only provide enough hydrogen to fill 18 vehicles per day but these are early days and it is promising to see a long-touted green technology finally getting off the ground.

A move away from the internal combustion engine to electric and hydrogen vehicles would clearly have positive effects on local air quality and, as such, should be encouraged. However, it is important to note that such a move would not completely eliminate all the sources of pollution associated with vehicle use. Firstly, even cars with zero exhaust emissions have air pollution – and other environmental impacts – associated with their full life cycle, from the initial resource extraction and vehicle manufacture through to final vehicle disposal. It may be that relatively little of the air pollution generated by such activities is emitted into the UK atmosphere but that is not the same as saying it is not produced at all; citizens of other, perhaps poorer, countries may be exposed to potentially harmful pollutants even if we are not. The second important point is that an appreciable amount of the PM pollution associated with vehicle movements arises from tyre wear, brake wear and road abrasion (Table 6.1); resuspension of road dust particles by moving vehicles is also a source of airborne PM. Therefore, the elimination of exhaust pipe emissions, whilst bringing massive air quality improvements, will not remove all air pollution associated with vehicle transport.

Fewer Vehicles

On even the most optimistic assumptions, universal usage of zero emission vehicles is some way in the future. Furthermore, even in the event of such a utopia coming to pass, it is clear that pollution would still be created, both locally and further afield, by vehicle

Table 6.1. Emission factors for PM_{10}[a] from cars. Source of data: NAEI (2012).

Emission factors	Petrol cars (g/km)	Diesel cars (g/km)
Cold start exhaust	Not estimated	0.065
Hot exhaust[b]	0.001	0.016-0.018
Tyre wear	0.007	0.007
Brake wear	0.007	0.007
Road abrasion	0.007	0.007

a. Figures for $PM_{2.5}$ are nearly identical for exhaust emissions and within the same order of magnitude (slightly lower) for other emission factors.
b. Range for urban, rural and motorway driving (no difference recorded for petrol cars across the different driving conditions, hence the single figure).

manufacture and use. It would make sense therefore to look for ways of minimising unnecessary travel and to encourage transportation methods that do not depend so heavily on the use of private vehicles.

Most proposals for reducing the number of vehicles on our roads focus on alternative means of transport. But before we investigate these, it is perhaps worth considering whether we always **need to travel** in the first place? In some cases, this may fundamentally be about a sense of entitlement to visit far-flung areas that our great-grandparents might not even have considered as a necessary or practical part of their lives. This relates to personal freedoms and the modern expectation that any of us should be able to go wherever we choose, as long as we can afford the trip. Many people are probably more familiar with a favourite exotic location, even if it is on the other side of the continent or the planet, than with the countryside and nearby towns in their own neighbourhood. On another level, there is the question of whether we now need to travel quite as often for business purposes as we once did. In the connected world of unlimited broadband, video conferencing and the like, it is not always vital for workers to be physically present in the workplace or all together at a particular meeting. Another factor influencing the 'need to travel' is town and country planning; the development of new suburbs and towns at commuting distance from major employment centres has long been an accepted part of planning policy since the railways were first built, long before car commuting even existed; but is it sustainable when air quality and wider environmental impacts are taken into account?

In an ideal world, more Britons would depend less on the car for local trips and more on **walking or cycling**. Both activities are good for the health and do not result in exhaust emissions to the local atmosphere. We do not live in an ideal world of course and, for better or worse, it is often seen as more convenient, and certainly easier physically, to hop into the car for that quick trip to the corner shop or the local take-away. Much is made of the need to 'educate' people of the benefits of physical activities like walking, but surely, most people already know that it is better for their health and for the environment; in fact, education of this kind is probably much less effective than the educators might hope, especially in a free society. It is certainly true that one person's pleasant twenty-minute stroll or five-minute bike ride to pick up a carton of milk is, to another person, something that would never be countenanced when a short, comfortable drive will do the same job. And, if it is raining hard, both people may be of the same (latter) opinion after all; furthermore, if the walk entails a steep climb along the way then fewer people are likely to be keen on walking-the-walk in the first place. As it is, the health and environmental benefits of walking and cycling might be best (or only?) encouraged by providing a pleasant and safe environment in which to undertake such activities. In recent decades, more and more cycle tracks have been opened in the UK and they tend to be well used if they are safe (usually meaning 'off-road'), well kept and in convenient locations. Similarly, people are more likely to prefer walking, into their local town centre for example, if they are not treated as second-class citizens. Too often, particularly in some of our urban areas, pedestrians are hemmed in to pavements by ubiquitous railings and made to wait at countless crossings whilst priority is given to through-traffic. This indicates that town and country planning is often focussed on motorists at the expense of pedestrians.

Walking and cycling may be good for the environment but it is inconceivable that they are ever going to remove (or even significantly reduce) the need for vehicle transport. In the case of private cars, one approach to reducing the number of vehicles on the roads is to make more efficient use of them. **Car sharing,** or 'pooling', has been encouraged in some countries, in the US for example, and can go some way to decreasing the total number of sole-occupancy vehicles making the daily commute. However, such schemes are unlikely to make a huge overall difference and, in any case, rely on both the wherewithal of individuals to be sufficiently organised and on their willingness to give up some personal autonomy and privacy, which are arguably among the most valued attributes of car ownership in the first place.

Potentially, the most effective way to get the public out of their cars is to provide them with a **public transport** network that can be genuinely described as safe, attractive, clean, reasonably-priced, convenient, reliable and punctual. The reduction in cars on the roads would clearly be beneficial for air quality but, ideally, the public transport vehicles should be zero-emission, entailing the use of electric vehicles or electrified transit systems using power provided by 'clean' electricity (see below). At the very least, public transport should increasingly become as non-polluting as possible in the short- to medium-term, using available and relatively-affordable technologies such as hybrid engines. Hybrid technology is already used in passenger transport vehicles (there are more than a thousand hybrid buses in London for example) and, whilst very few buses in the UK are fully electric at present, Transport for London aims for all single-decker buses to be electric- or hydrogen-powered by 2020 (TfL, 2015). Looking further into the future, such plans will need to spread out from the capital to Britain's other towns and cities and even to rural public transport networks.

Providing public transport that meets all of the criteria listed above is easier said than done and would require significant investment. The total annual subsidy for public transport in Great Britain already stands at around £7.5 bn, with some two-thirds of this going to the railways and the remainder to subsidise bus services, mainly through concessionary fares[3]. Increasing subsidies further would mean higher taxes, or diversion of public money from other needy causes, but would bring, in economic terms, external benefits. Such benefits would include less congestion and pollution on the roads and a healthier and more connected society.

Arguably, the biggest challenge, practically and financially, for improving public transport is to match, or at least approach, the convenience of a private vehicle. For those living on bus routes or near to railway and metro stations, such as those in London and other large cities, public transport may already be more convenient than driving, at least within cities with some sort of integrated public transport policy (i.e. where different routes link to each other). But for everyone else, such convenience would require large-scale investment, together with sufficient public demand and the necessary political will. In any case, even a fully integrated public transport system may not serve every purpose in modern Britain, precisely because of cultural changes that the car itself has made possible. For example, for many years now the car has enabled many Britons to 'escape to the country', to live the rural idyll deep in the countryside, whilst still commuting daily to their urban workplace. A public transport service that would allow everyone, including those living in remote hamlets, to leave the car at home and come and go as they please would be a significant challenge.

Another example of cultural change brought about by widespread car ownership is the tendency of many households to do a weekly shop at an out-of-town supermarket, taking home a whole car-load of provisions. Most people would probably baulk at the idea of carrying numerous bulging shopping bags all the way home from the nearest shops, even if public transport is available for all or part of the journey. An interesting development in this context has been the growth in recent years of online grocery shopping, where the weekly shop is ordered via a supermarket website and brought to the house in a delivery van. This has been enabled by the wider availability of the internet, just as out-of-town shopping was enabled by increased car ownership in the late-twentieth century. It is conceivable that online shopping could, to some extent, reverse the polluting effects of out-of-town supermarkets, particularly if the delivery vehicles were powered by renewable electricity or hydrogen. It is unlikely to be a panacea however; on the one hand, it could mitigate against improved public transport in the way car ownership already has, by fragmenting the demand and 'critical mass' that a fully integrated public transport system requires. It could also increase vehicle movements if, for example, items of clothing that would once have been purchased by one person on a single shopping trip have to be delivered by several vehicles, all from different online stores.

Despite the real challenges to public transport posed by the investments that would be required and the apparent public preference for private transport, a truly modern public transport system that served the greatest number of people that circumstances and finances allowed, and fulfilled all the criteria mentioned above, would be likely to encourage more people out of their cars and reduce the number of polluting vehicles on the road, bringing benefits to society as a whole, particularly in the form of improved health and life expectancy.

An additional approach to the problem of traffic pollution is to remove the most polluting vehicles from our roads by: (i) discouraging their use in general, using financial instruments and; (ii) prohibiting them entirely in the most badly-affected areas. One of the most widespread **economic tools** is the vehicle tax banding system described in the previous section, whereby the oldest, most polluting cars attract much higher levels of tax, effectively discouraging their continued use. Another is the tax on petrol and diesel; whilst this tax is paid by everyone (except those using electric or hydrogen cars), those driving the most fuel inefficient cars are hit the hardest so, again, the fuel tax theoretically discourages the use of the most polluting cars. A more focused financial disincentive to using polluting vehicles is the congestion charge, as levied for some years now in central London where the daily charge is £11.50; the charge is waived for 'plug-in' electric vehicles and for conventional cars and vans meeting the Euro 5 emissions standards (Table 6.2). Congestion charging is rare in the UK with only one other example, in Durham. A planned charging zone for Manchester was dropped after the public voted four-to-one against the proposal in a referendum and similar rejections have occurred in Edinburgh and Birmingham.

Table 6.2. Euro standards for private cars[a]. Units are grams of pollutant emitted per kilometre of travel (g/km).

		Petrol		*Diesel*	
Standard	Year	NOx (g/km)	PM (g/km)	NOx (g/km)	PM (g/km)
Euro 1	1992	-	-	-	0.14
Euro 2	1996	-	-	-	0.08
Euro 3	2000	0.15	-	0.56	0.05
Euro 4	2005	0.08	-	0.30	0.025
Euro 5	2009	0.06	0.005	0.23	0.005
Euro 6	2014	0.06	0.005	0.17	0.005

a. Separate standards (not shown) have been set for buses and lorries in each of the years shown in the table, based on emissions per kilowatt hour of energy output. These have been similarly tightened with each new set of standards.

Outright prohibition of the most polluting vehicles from affected areas has generally been achieved in the UK, and elsewhere, by declaring '**low emission zones**' (LEZs)[4]. Currently, these prohibitions generally apply to heavy goods vehicles (HGVs) and buses rather than passenger cars. In London, only buses, lorries and large vans that meet proscribed Euro standards are allowed to enter the city's LEZ, which, unlike the central London congestion charge, covers most of Greater London[5]. From 2020, an 'ultra low emission zone' in London will apply to cars as well. Elsewhere in the UK there are LEZs, in Brighton, Norwich, Nottingham and Oxford, which apply to buses only – again only vehicles meeting specific Euro emissions standards are permitted to enter.

Another form of vehicle ban is the '**car rationing**' that authorities sometimes impose temporarily during severe pollution episodes. Such bans are typically applied on alternate days to vehicles with odd- or even-numbered registration plates so that the amount of traffic on any given day is effectively halved during the pollution episode. This approach has been applied in polluted cities around the world including Athens, Beijing, Bogota, Delhi, Mexico City, Paris, Santiago and Sao Paulo, although never in the UK. In some cases – for instance in Milan in 2015 – air pollution episodes have been so severe that all vehicles have been banned for short periods of time.

A major form of road transport that we have not yet considered (except briefly in the London LEZ) is **freight and haulage**. Heavy goods vehicles transport essential goods into, out of and around the country. New HGVs must meet the current Euro emissions standards and, as such, emit less pollution than older vehicles; however, HGVs are expensive and are generally used for many years so most of the road freight we depend on is currently transported by relatively polluting diesel vehicles. As emissions standards progressively tighten, the situation will improve to some extent and in the longer term, zero emission (e.g. electric) HGVs may become more viable than at present; however, for the foreseeable future, HGVs are likely to remain important sources of air pollution. Shifting freight to the railways, which could conceivably be run on renewable or nuclear power, is sometimes suggested but is unlikely to provide the full answer, mainly because of the greatly increased rail capacity that would be required.

Reducing Pollution from Power Generation

In post-industrial Britain, the generation of electricity in fossil fuel-fired power stations is a major contributor to air pollution. Compared to transport sources, the emissions are more remote from human populations but the pollutant volumes are large and contribute significantly to atmospheric pollution. Therefore, looking for ways to reduce pollution from power generation is vital; even more so if we accept that the vehicles of the future will need to be powered, directly or indirectly, by clean electricity[6].

Reaching zero emissions from the power generation process will require a wholesale shift away from fossil fuel combustion. Whilst some institutions, including government bodies, universities, religious organisations and pension funds, have already engaged with divestment from fossil fuels, the global economy is still strongly dependent on oil, coal and gas extraction. Over the centuries, the UK has utilised its vast reserves of coal as a ready source of energy; however, the last mine closed in 2015, bringing centuries of deep coal mining to an end. British coal-fired power stations will continue for some years yet, burning imported coal, but the government has proposed to close them down by 2025. This is part of the '**dash for gas**', a shift to gas-fired power stations that started in the 1990s and has been partly driven by the UK's commitments to reduce carbon emissions, because gas emits less carbon dioxide than coal per unit of energy produced. However, natural gas is, of course, a fossil fuel and burning it still produces carbon dioxide and other pollutants including NO_2 and unburnt hydrocarbons. The UK government is looking towards shale gas as a future source of energy but this will also be a source of air pollutants and there are other environmental concerns associated with the extraction of shale gas, via the technique of hydraulic fracturing or 'fracking'.

Generating power by burning natural gas or shale gas will inevitably cause air pollution and other sources of power generation will be required unless we are prepared to accept continued poor air quality and the health problems it brings. There are two broad alternatives for generating power without producing air pollution, at least during normal operation; one of these is renewable energy and the second is nuclear power.

Renewable energy depends on capturing the inherent power of the sun, wind, tides and other sources such as ground heat. Public support for renewables is consistently high in the UK at 75-80%[8] (DECC, 2015b) yet local councils are increasingly rejecting applications for wind and solar projects; for example, in 2014, more than half of all UK wind farm applications were rejected (Tait, 2015). Two of the main arguments against the uptake of renewables are that they are intermittent, unreliable sources of energy and that they are unsightly blots on the landscape. A counter argument to the first objection is that they should be seen as part of an 'energy mix', whereby different sources of energy are tapped into at different times. In fact, renewable energy sources were providing nearly a quarter of the UK's electricity in mid-2015[7](DECC, 2015a), although just under half of this was from biomass combustion, which generates air pollution[9]. Wind power is certainly intermittent but progress is being made towards better batteries for storing the captured energy. Solar panels do not need constant, direct sunlight and even in relatively cloudy places like the UK, they could massively reduce our dependence on other power sources.

Concerns about solar panels and wind turbines blighting the landscape are subjective; many people actually like to see wind turbines[10] and, arguably, they are no more intrusive than the electricity pylons that have been part of the British landscape for many decades. Turning over attractive farmland to arrays of solar panels may not be universally popular but, given the right support, solar panels could instead be fitted to the roofs of houses and buildings, providing abundant clean energy with relatively little visual impact.

Renewables are also criticised by opponents for requiring subsidies that increase consumers' energy bills, via 'green levies'. The Conservative government elected in 2015 was at the forefront of this argument, having made a manifesto commitment to remove public subsidy for onshore wind power. The UK's Department of Energy and Climate Change stated that subsidies for renewable energy sources in 2014-15 amounted to £3.54 bn (DECC, 2015c) and the government subsequently reduced subsidies for renewables. For example, as well cutting support for onshore wind and small-scale solar power under the 'Renewables Obligation', it reduced the subsidised 'feed-in tariff' for solar power. These moves were widely criticised as hitting the fledgling renewables industry before it had become fully established, just at a time when it had been making excellent progress.

Meanwhile, financial support for polluting fossil fuel energy continues, via subsidies and, in the words of the aforementioned Conservative manifesto, via 'tax cuts [that] have encouraged record levels of investment in existing North Sea gas, and the birth of a new industry, shale gas' (Conservatives, 2015). In the context of subsidies, the International Energy Agency states that, globally, subsidies for fossil fuels are six times higher than for renewables (IEA, 2013)[11]. The International Monetary Fund similarly noted that global fossil fuel use is massively subsidised, especially when its environmental costs are included (IMF, 2015). The IMF's estimate of total global subsidies for fossil fuel use, based on 2013 figures, is a massive $4.9 trillion, amounting to 6.5% of global GDP. It estimates that, on average, each UK citizen pays $635 (about £418) per year amounting to a total UK subsidy of $36.7 bn (approximately £24 bn). The IMF states further that eliminating fossil fuel subsidies would more than halve global deaths from air pollution and raise $2.9 trillion of revenue worldwide. Despite these figures, and contrary to its rhetoric on the international stage about the need to cut greenhouse gas emissions, the UK government that was elected in 2015 pressed ahead with swingeing cuts to renewables whilst continuing its support for polluting fossil fuels.

Nuclear power, like renewable energy, can provide power without polluting the air during normal operation; however, it is a technology that strongly divides opinions, even amongst environmentalists. The main arguments against nuclear energy are that it is expensive, generates long-term, highly radioactive waste and puts populations at risk from radiation releases following accidents, such as those at Chernobyl in 1986 and Fukushima in 2011. An indication of the high costs of nuclear power is provided by the case of the Hinkley Point C reactor in Somerset, which is to be built by 2025 at a cost of £18 bn. The huge investment required by the operators, EDF of France and CNG of China, was agreed by the UK government, which guaranteed a price for the electricity produced of £92.50 per megawatt hour; this was more than double the wholesale price of electricity at the time and the agreement was widely criticised as amounting to a potential levy on consumers' future bills, far in excess of the green levies associated with renewable energies.

In the context of radioactive waste, each year, a 1000 megawatt nuclear power station produces approximately 30 t of the most hazardous, 'high-level' waste (IAEA, 2014), which remains highly radioactive for hundreds to thousands of years. The safest management option is to vitrify the waste and store it deep underground but populations in proposed burial areas have strongly opposed the plans and high level waste is currently stored above ground in water tanks or dry-cask storage vessels. The fear of nuclear accidents is understandable, despite the loss of life from such accidents being, in fact, much less than that caused by the burning of fossil fuels. Radiation from accidents, such as the one at Chernobyl, should also be put into context with other sources of radioactivity, as shown in Table 6.3. Despite the counter-arguments that can be put forward for nuclear power, the concept of risk is particularly difficult to convey and public opposition is likely to remain strong.

Table 6.3. Radiation doses (mSv) in context. Adapted from Peplow (2011).

Dose	Source
0.3	Total dose received by each resident of Europe for 20 years after Chernobyl
2.4	Average annual background radiation globally
9	Total dose received by the 6 mn residents in Chernobyl-contaminated areas (>37 kBq m^{-2}) of the former USSR
9	One computed-tomography (CT) scan
9	Annual exposure of airline crew flying regularly between New York and Tokyo
30	Average total dose of external radiation received by evacuees from Chernobyl plant and surrounding area
120	Average total dose received by liquidators at Chernobyl (1986–90)

If the UK is to generate its power requirements without further polluting the atmosphere, could this be accomplished using only renewable energy? In theory, the answer is yes, particularly if advances in energy storage technologies continue, but the public's strong backing of renewables would need to be matched by commensurate government support, which is sadly lacking at the moment. Another important aspect in this context is **energy efficiency**. Too much of the power available to us is simply wasted, in over-heated or poorly insulated buildings for example, or in brightly lit but empty rooms. A real commitment to energy efficiency is required and this needs to come from individuals, institutions and companies, as well as from central government. Unfortunately, official policies often run counter to the adoption of energy efficiency measures. For example, a long-standing commitment for new homes in the UK to have zero carbon emissions was cancelled by the government and an initiative called the Energy Company Obligation scheme, whereby energy companies provide homes with energy efficiency measures, was significantly scaled back. Another example is the 'Green Deal', which gave financial incentives to householders to improve the energy efficiency of their homes, but was also scrapped.

Adopting stringent energy efficiency measures would not just contribute to cleaner air in a theoretical future but could help to reduce air pollution today because it would mean less burning of the fossil fuels we currently depend upon. In the meantime, the main option for limiting pollutant emissions from our current fossil fuel-fired power stations is to use the best available **pollution control technology** (PCT), such as low NOx burners and flue gas desulfurisation, which reduce NO_2 and SO_2 emissions, respectively. However, PCT is a reactive, rather than a proactive measure and, whilst it reduces atmospheric pollution, it by no means eliminates it. As we have seen, achieving the latter would require the wholesale adoption of renewable energy and a real commitment to energy efficiency, possibly together with a continuing contribution from nuclear power if its safety risks were at a publicly-acceptable level.

Reducing Pollution in other Ways

Clearly, transport and power generation are major sources of Britain's air pollution and should be targeted as a priority; but there are other sources that can be particularly important on a local level and should not be ignored when considering approaches to improving air quality.

Arguably, the most important, but perhaps the most difficult to address, is **farming**. Agricultural emissions of nitrogenous gases, particularly ammonia, combine with oxides of nitrogen and sulfur in urban and industrial areas to form fine PM, significantly decreasing air quality. Reducing ammonia emissions from farming can be achieved; for example, by incorporating manure into the soil at a shallow depth rather than spreading it onto the surface. Another approach is to limit the use of synthetic urea fertilisers, which are an important source of ammonia, or to incorporate urease inhibitors into the soil to reduce the natural conversion of urea to ammonia. However, such novel approaches would need widespread changes to age-old, perhaps entrenched, practices to be effective and there is the concern that any reduction of fertiliser use (synthetic or organic) could reduce crop yields and food security. Furthermore, the continual yearly increases in worldwide meat consumption suggest that livestock farming is likely to become an ever-larger source of ammonia emissions for the foreseeable future. It is important to remember though that agricultural air pollutants generally react with urban and industrial emissions to cause the worst problems of poor air quality and, encouragingly, studies have indicated that reducing pollutants from combustion sources, as detailed in the previous sections, could be the key solution; for example, reductions in combustion NOx alone are predicted to reduce $PM_{2.5}$ concentrations by a similar margin to that obtained when both agricultural ammonia and combustion NOx are reduced (Bauer et al., 2016).

We saw in earlier chapters that the main cause of the Great Smog of 1952 in London was **domestic fires**. The subsequent introduction of smoke control areas and the widespread ownership of central heating systems that also followed, means that domestic fires now comprise a relatively minor contribution to poor air quality when compared to their role in the worst smogs of the past; however, there is still cause for concern. Firstly, there is no ready method of finding out where smoke control areas are in the UK and, therefore, whether it is legal to light a domestic fire in a particular area. Information does not appear to be routinely supplied to householders, and government websites merely refer those interested to their local council. Therefore, unless every householder in the UK phones up their local council to ask, it is likely that many Britons will remain blissfully unaware of whether or not they live in a smoke control area. This seems to throw up unnecessary obstacles and surely increases the risks of unsuspecting householders lighting domestic fires in locations that are officially, but quite anonymously, smoke control areas.

The main concern relating to domestic fires is that outside smoke control areas (and, considering the point above, very possibly within them too), the pollution emitted can accumulate at ground level in certain conditions, greatly increasing the risks of human exposure to fine PM and other pollutants. During cold spells, the likelihood of pollution from domestic fireplaces and wood burners increases as people understandably try to keep warm; but this happens just as temperature inversions are more likely to be forming, keeping the smoke at ground level. Whilst the worst of the 'great smogs' of the past are not likely to be repeated in Britain today, a neighbourhood with a number of smoking chimneys on a cold winter's night can still create a localised (but probably unrecorded) pollution episode.

With these concerns in mind, what can be done? As a minimum, there should be clear information given to all householders living within smoke control areas about the controls applicable to them. Further, to ensure that pollution from domestic fires does not accumulate to dangerous levels in cold spells, the UK could adopt initiatives like the 'Spare the Air' scheme in the San Francisco Bay Area of California. Here, solid fuel burning in fireplaces or wood burners is declared illegal during winter 'alert' periods of cold, stable air. On such 'no burn days', residents have to wrap up warmly or use only central heating or plug-in heaters so that no smoke is emitted in the affected area. Whilst the UK's introduction of smoke control areas in the 1950s undoubtedly improved a drastic situation, the use of wood-burners and the burning of solid fuels such as anthracite coal in open fireplaces is still permitted, even within the designated smokeless zones; perhaps a stricter approach along the lines of 'Spare the Air' would avoid the localised, but potentially serious, pollution that domestic fires can still bring.

Industry in Britain declined rapidly from the 1980s onwards but, locally, industrial sources of air pollution still require careful monitoring and management. Industrial emissions are regulated by the system of Integrated Pollution Prevention and Control (IPPC), which covers not only power generation and traditional industries like metal processing, chemical production and food and drink processing but also other potentially polluting sources such as waste treatment facilities and intensive livestock installations. Industries regulated by IPPC must obtain a permit to operate and are required to use the 'best available techniques' for controlling air pollutants. Emissions management is generally focussed on PCT, as discussed in the previous section in the context of power generation. Often, PCT takes the form of pollutant removal from the flue gases but, in some cases, other approaches are employed. For example, modern waste incinerators in the UK are generally operated at sufficiently high temperatures to minimise the formation of dioxin, a particularly toxic compound that was emitted in much greater quantities from earlier, mass-burn incinerators that did not always have the same level of fine temperature control.

A major section of this chapter has focussed on reducing emissions from road vehicles but other forms of transport also need to be addressed. In the context of **air travel**, most pollutant emissions from aeroplane engines occur at altitude; this raises concerns about climate change but has no direct impact on local air quality at ground level. However, emissions from aeroplanes that are taxiing and taking off and landing are of more concern[12], as are emissions from airport site traffic and, in particular, pollution from vehicles travelling to and from airports. For these reasons, it follows that a reduction in air travel would lead to improved air quality in locally-affected areas. Clearly though, such a reduction is unlikely to occur in the foreseeable future and may well not be wanted by the average Briton, regardless of the direct and indirect contributions of air travel to atmospheric pollution and ill-health. Many Britons are now well accustomed to the personal freedom of being able to travel anywhere in the world as long as they have the financial means. But should we comfortable with ongoing increases in airport capacity, in the knowledge of the environmental effects that air travel has? The thorny issue of expanding London's airport capacity has been debated for some years, partly on air quality grounds, but is likely to happen in some form. A proposal for a third runway at Heathrow was backed by the Airports Commission in 2015 but the House of Commons Environmental Audit Select Committee said the government should not approve expansion at Heathrow unless it could show that legal pollution limits would be met (HCEAC, 2015).

Whilst most international travel is by aeroplane, the main mode of freight transport around the world is **shipping**, added to by large numbers of passenger ferries. Just as aeroplanes emit much of their pollution at altitude, emissions from ship and ferry engines occur mainly at sea, similarly remote from human populations, and traditionally shipping has been one of the least regulated sources of air pollution. However, there are tens of thousands of sea-going vessels worldwide, many powered by polluting, sulfurous, heavy fuel oil, and at any one time thousands of ships and ferries around the world are at dockside, generating pollution in the vicinity of human populations. This is added to by long-range transport of pollutants generated off-shore. Brandt et al. (2012) estimate that in Europe alone, air pollution from shipping contributes to approximately 50,000 'premature deaths' each year. Measures that could be taken to reduce pollution from shipping include: (i) the use of shore-side electricity when in dock; (ii) greater fuel efficiency (mainly by reducing the speed of travel); (iii) the use of low sulfur fuels or liquid natural gas and; (iv) engine modifications and exhaust catalysts to reduce NO_2 emissions[13]. Vessels operating in the North Sea, Baltic Sea and English Channel (designated as 'sulfur emission control areas') are now required by EU law to use fuels with a sulfur content of 0.1% or less. Whilst this will reduce the pollution and ill-health associated with sulfur dioxide in these areas, a higher level of sulfur in fuel is allowed in other European waters and action is still required in relation to other serious air pollutants from shipping, particularly NO_2 and fine PM.

A number of actions, initiatives and approaches have been suggested in this chapter as potential solutions in the pursuit of a cleaner atmosphere; they are mostly proactive, with a focus on preventing or minimising air pollution arising in the first place. However, for the foreseeable future, poor air quality is likely to remain a reality, particularly during pollution episodes; so, what **reactive approaches** could be taken to remove noxious gases and particles that are already present in the air we breathe? Some novel solutions are proposed, although not all have been totally successful, highlighting the importance of proactive measures to prevent the initial generation of pollution. For example, between 2010 and 2012, Transport for London piloted the application of calcium magnesium acetate (CMA) dust suppressants to London road surfaces in an effort to reduce airborne PM_{10} concentrations; CMA forms a hygroscopic coating on the treated surface keeping it damp and so reducing resuspension of dust. Some success (up to 44% reduction in local PM_{10} concentration) was observed around industrial and construction sites, where site vehicles typically resuspend on-site surface dust, but no significant effects were noted at other roadside locations (Barratt et al., 2012). Researchers at Sheffield University have put forward interesting plans to use nanoparticles of harmless titanium dioxide in roadside billboards and even in clothing to remove nitrogen oxides and volatile organic compounds from the air but, so far, the initiative has not been widely adopted. Another suggested approach is to plant trees by the sides of busy roads and creepers like ivy on roadside walls ('green walls') to capture some of the PM and NO_2 emitted by vehicle exhausts, preventing it from reaching pedestrians and householders. This would be a relatively low cost option and, most probably, a popular approach but care is needed to select the most appropriate and robust species and ensure that tree canopies do not actually have the unintended effect of trapping pollutants at ground level.

CHAPTER SEVEN

Conclusion: Looking Forward to Clean Air for All

Cleaning up our polluted atmosphere is vital if we are to improve our health and life expectancy and protect our most vulnerable loved ones; our children, our sick and our elderly relatives. Achieving clean air will not be easy and will not happen overnight; it will require wholesale changes to the way we do things and not all of the changes will be universally popular. But it is by no means an insurmountable challenge and we must surely confront it head on. We now have a clearer understanding than ever before of the extent and impacts of air pollution in the UK, together with the approaches and clean technologies that can reduce it or even eliminate it completely.

Realistically, our best hope of rapidly reducing emissions from road transport, the source of Britain's worst forms of air pollution, may lie in the further development and increased support and take-up of zero emission vehicles. Serious measures to *reduce* vehicle numbers are, in all reality, unlikely to be sanctioned by governments because a government in power will always be eager to please its electorate; and the truth is that the average UK voter shows no signs of wanting to exchange his or her car any time soon, for alternatives like public transport or walking. This might change to some extent if British citizens were made fully aware of the severe impacts that current levels of air pollution have on our health but there is no certainty that it would lead to the population-scale behavioural changes that would be required.

If technology is to be instrumental in addressing poor air quality, we should take heart from the fact that hybrid vehicles, which are a useful stepping-stone to achieving zero-emission, are already common. Looking ahead, fully electric vehicles are already becoming more affordable and even the long-touted hydrogen-fuelled car is now being commercially produced for the first time. Notable in these respects is the bold aim of the leading political parties in Norway to ban the sale of fossil fuel-powered vehicles from 2025. It will be interesting to see how this progresses and whether other governments will follow suit.

To be truly emission-free, the power supplied to electric vehicles, and required to supply hydrogen for fuel cells, will need to be generated by non-polluting means, as will our other power needs. With continuing public distrust of nuclear power, renewables like solar and wind are likely to be the preferred option. But they will only be able to provide all our energy needs if they have the full support of sufficiently concerned and motivated governments, wedded to a strong commitment to energy efficiency.

World leaders have, belatedly, come to an agreement of sorts on the need to confront the great environmental issue of our generation, human-forced climate change[1]. Any moves to decarbonise the world's economy would, eventually, have the additional effect of improving local air quality as we gradually move away from burning fossil fuels. But there is a problem. The timescale discussed for decarbonisation is measured in decades and focuses on limiting the extent of global warming by the end of the century: Poor air quality is a public health crisis *now* and action is required urgently.

The deaths of thousands of Londoners in the winter of 1952-53 galvanised public and political opinion sufficiently to act on air pollution, culminating in the Clean Air Acts and putting an end to the deadly 'pea-souper' smogs of history. We now know that a less visible form of air pollution is still causing serious respiratory diseases in the twenty-first century and shortening life expectancies across the UK, and the world, every year. It is surely time to put a stop to this by considering our actions, by acting responsibly and by campaigning for change.

EPILOGUE

A Sting in the Tale?

Despite the problems caused by air pollution and the importance of doing all we can to reduce its presence in the air, there is an unintended consequence of PM pollution in mitigating another, more well-known environmental impact. Some of the fine particles that human activities emit to the atmosphere reflect incoming sunlight away from the Earth, leaving the Earth's surface and the surrounding atmosphere cooler than it would otherwise be. Other pollutant particles are composed mainly of soot and, being black, absorb sunlight and conduct the absorbed energy as heat into the surrounding atmosphere, warming it slightly. Atmospheric scientists have established that the reflective PM cooling exceeds the atmospheric heating effect of black soot particles. Interestingly, what this means is that human PM emissions, such an important factor in our dangerously poor air quality, are actually masking some of the effects of global warming. And it follows that as we, quite rightly, take the necessary measures to improve air quality, the masking effect will be removed and the Earth may warm more quickly than had been expected. Climate scientists predict that the rate of global warming could increase from 0.2 °C per decade to 0.8 °C per decade in the northern hemisphere because of this effect (Raes and Seinfeld, 2009).

None of this means that we should slacken our efforts to remove PM and other pollutants from the air (or even inject reflective sulfate particles into the atmosphere, as has been suggested as a technical 'fix' for global warming); rather it tells us that we should attack atmospheric pollution on all fronts – the greenhouse gases that are warming our planet but also the less well-known air pollutants, such as PM, NO_2 and O_3, that are harming our health and causing excess mortality equivalent to tens of thousands of deaths in the UK and millions around the world.

APPENDIX ONE

Air Quality Standards

Exceedances of UK Air Quality Standards, 2011-15. Based on data available from DEFRA (2016b).

	AURN monitoring sites exceeding UK Air Quality Standards [a]			
	PM_{10}[b]	NO_2	NO_2	O_3
Year	Daily Mean	Hourly Mean	Annual Mean	8-hour Mean
2011	90% (55/61)	14% (17/118)	13% (15/118)	94% (77/82)
2012	95% (59/62)	18% (22/121)	17% (20/121)	93% (78/84)
2013	91% (59/65)	15% (18/118)	14% (16/118)	92% (77/84)
2014	84% (55/65)	14% (17/124)	15% (19/124)	89% (71/80)
2015	88% (60/68)	13% (17/132)	14% (18/132)	95% (75/79)

a. Shown in brackets are the numbers of monitoring stations exceeding the AQS on one or occasions that year, out of the total number of UK stations monitoring for the pollutant in the same year.

b. The table does not include $PM_{2.5}$, which is of greater health concern than PM_{10}, because a binding limit value had not yet been set for it during this period.

APPENDIX TWO

The estimation of deaths attributed to air pollution

For a brief summary of the following, see also note 1, Chapter One *(page 94).*

An estimated annual number of UK deaths attributed to air pollution on the population scale was calculated by the Committee on the Medical Effects of Air Pollutants (COMEAP, 2010). The estimates were calculated using a concept called the risk ratio (RR), a measure of excess mortality from a suspected cause, over and above the mortality expected from all other causes. The RR used, selected from previous detailed epidemiological research of hundreds of thousands of subjects, was 1.06 for long-term exposure to a $PM_{2.5}$ concentration of 10 $\mu g\ m^{-3}$; this translates to an estimated excess mortality rate at that concentration of 5.7%. This RR was adjusted for modelled $PM_{2.5}$ concentrations in each UK local authority area[1] and was applied to the total number of deaths in each local authority area (as described in Gowers et al., 2014). The calculated estimates of deaths attributable to anthropogenic $PM_{2.5}$ pollution in each local authority area were then totalled to give regional figures and, ultimately, the national figure of 28,969 attributable deaths (Gowers et al., 2014). The average amount of life lost for each attributable death, which varies with the different life expectancy of each age group in the adult population, was estimated at twelve years.

A number of uncertainties and assumptions are included in the estimations and these are detailed in the COMEAP report and the later report by Gowers et al. (2014), which states:

> "...long-term exposure to air pollution is understood to be a contributory factor to deaths from respiratory and, particularly, cardiovascular disease, i.e. unlikely to be the sole cause of deaths of individuals"

and

> "...it is likely that air pollution contributes a small amount to the deaths of a larger number of exposed individuals rather than being solely responsible for a number of deaths equivalent to the calculated figure of 'attributable deaths'."

In a similar vein, COMEAP (2010) emphasises that the calculations relate to population estimates and states:

> "...it is likely that exposure to $PM_{2.5}$ acts in conjunction with other risk factors to cause earlier death..."

and

> "...air pollution has the potential to affect everyone who breathes the air. The effects are principally on mortality from [...] cardiorespiratory causes [...] and from lung cancer. These are complex diseases, with multiple established and likely causes at the population level [...]. Given this complexity, it is not plausible to think of 'attributable' deaths as enumerating an actual group of individuals whose death is attributable to air pollution alone."

With such considerations in mind, COMEAP (2012) recommends that calculations of excess mortality are expressed as 'an effect on mortality equivalent to 'X' deaths at typical ages' and also uses the term, 'attributable deaths'. In reference to communicating the mortality estimates, COMEAP (2010) states:

> "We stress the need for careful interpretation of these metrics to avoid incorrect inferences being drawn – they are valid representations of population aggregate or average effects, but they can be misleading when interpreted as reflecting the experience of individuals."

Bold statements in newspaper reports that, for example, 'air pollution kills 29,000 people a year' are, strictly speaking, inaccurate, although they do convey an important message (and to a wide readership) that air pollution in the UK is a major cause of serious ill-health and a significant source of excess mortality, as represented by the calculation of attributable deaths.

In summary, our polluted atmosphere is, without doubt, 'an air that kills' but the mortality estimates should be considered, as intended, in the context of loss of life expectancy on a population scale, or as 'attributable deaths' rather than a number of deaths of specific, identifiable individuals.

Notes

Chapter One

1. Media outlets sometimes report that a certain number of people die from air pollution each year in the UK or elsewhere. However, it should be understood that these reports are actually based on estimates of deaths on a population scale; i.e. a number of deaths representing the likely level of overall excess mortality in a population caused by mass exposure to air pollution. This is not the same as saying that, each year, a certain number of individual people can be identified as having died from exposure to air pollution. This is because it is not possible to say, for example, that someone who died from respiratory disease did so only because of exposure to air pollution. With this in mind, a more accurate way of reporting the estimates would be to state that air pollution has an effect on mortality, on a population scale, which is equivalent to the stated number of deaths. For further information, the reader is urged to consult the details about mortality calculations that are included in Appendix 2.
2. The total number of deaths attributed to air pollution globally is estimated by the World Health Organisation at 7 million, but 3.3 million of these are attributed to exposure to indoor air pollution, mainly from open cooking fires in developing countries (WHO, 2014a).
3. The three most common gases in dry air are nitrogen (78%), oxygen (21%) and argon (0.9%). As humidity increases water vapour becomes an important component too, comprising up to 4% of the atmosphere.
4. The increase in atmospheric oxygen at this time would have been catastrophic for the world's anaerobic organisms – to them oxygen is a poisonous pollutant and such life-forms survive today only in oxygen-free environments like pond slime and mammal intestines. Even aerobes, which require

oxygen to survive, have had to evolve antioxidant defence mechanisms to offer some protection against cellular oxygen toxicity.

5. Nitrogen can also be fixed naturally by lightning bolts, which provide the energy needed to break the very powerful chemical bond that binds two nitrogen atoms together in a gaseous N_2 molecule; microorganisms similarly fix N by breaking this bond. This enables the formation of nitrogen compounds that can be rained out of the air to soils and which, unlike N_2, can be used to build amino acids. For the last 100 years, atmospheric nitrogen has also been fixed by the industrial 'Haber' process, mainly for incorporation into fertilisers.
6. The term 'pollution' is applied when the contaminants of a medium like air or water are of a nature and/or quantity that can cause harmful effects.
7. Other countries in this category include Turkey, South Africa, Thailand, Malaysia, Indonesia and the Philippines.

Chapter Two

1. For anyone interested in the historical aspects of environmental issues and/or the history of London, *The Big Smoke* (1987) by Peter Brimblecombe is an excellent read; entertaining and meticulously researched. Since its publication, much of what has been written on the early history of air pollution in England appears to be referenced to this book (or can be traced back to it). The first paragraph of the current chapter is also informed by it to a large extent.
2. The widespread presence of near-surface coal deposits in the area is another contender as the source of the name; perhaps the blackness underfoot but also everywhere else, once the abundant coal had been burned, explains the use of the name.
3. The word 'smog' is usually attributed to Dr Henry Antoine des Voeux who used it at the 1905 Public Health Congress in

London but he seems to have indicated that the word was already in general use, as a contraction of 'smoky fog'.

4. The meaning of this term is explained towards the end of Chapter 3.
5. These included other documented episodes that also caused, collectively, the deaths of thousands of people.
6. This was clearly nothing new; Sheffield's houses were described in 1724 by Daniel Defoe as 'dark and black, occasioned by the continued smoke of the forges' (*A Tour through the Whole Island of Great Britain*), whilst William Cobbett in his *Rural Rides* of 1830 states, 'They call it Black Sheffield, and black enough it is'.

Chapter Three

1. In practice, it is likely that emissions also occur within smoke control areas (see Chapter 6).
2. Ultrafine particulate matter known as $PM_{2.5}$ (defined in the next section of this chapter) was used to derive the modelled estimates because this one pollutant is stated by the study authors to capture around 80% of the health impacts of air pollutants, partly because it is a key indicator of combustion emissions. The authors acknowledge the uncertainties in the modelled estimates but warn that the impacts of road traffic are likely to have been underestimated in the outputs because of the inherent toxicity of particles associated with vehicle emissions and the limited resolution of the model in representing short-term peaks in roadside pollutant concentrations.
3. Despite the decision, in a 2016 referendum, of the British people to leave the EU, the UK's legal, EU-set commitments on air pollution will remain, at least until such time as the country formally exits the Union (a process that takes up to two years from formal declaration of Article 50 of the Lisbon Treaty). Furthermore, upon exiting the EU, the UK government will remain under pressure to retain the existing

AQS concentrations, regardless of EU membership – not least because the AQSs are, in turn, based on WHO guidelines.

4. In this book, discussions of air pollution impacts are focussed on direct human health effects but there are other serious impacts, particularly on wild and cultivated plants, on various materials (e.g. rubber, which is damaged by ozone) and on visibility. Also, nitrogen dioxide and sulfur dioxide are both acidifying substances, able to cause 'acid rain'.
5. For example, gaseous ammonia combines with gaseous nitrogen oxides to produce tiny ammonium nitrate particles which can comprise significant percentages of the $PM_{2.5}$ and PM_{10} fractions, particularly in spring when mineral and organic fertilisers are spread on farmland and ammonia emissions peak. Livestock agriculture is the main source of ammonia and, as such, can be a major contributor to PM pollution. In southern California, where air pollution has been studied for many decades in and around Los Angeles, ammonia from upwind livestock areas is thought to contribute significantly to the high concentrations of $PM_{2.5}$ suspended in the atmosphere (Hasheminassab et al., 2014).
6. Also, because they are small enough to remain suspended in the air for weeks and travel long distances, often crossing national boundaries.
7. In fact, this process first forms nitric oxide, a simpler molecule containing just one nitrogen atom and one oxygen atom (NO); much of the NO is then quickly oxidised in the air to NO_2.
8. Similarly, the term ozone hole, whilst signalling an important problem, is also something of a misnomer as it does not relate to zero concentrations of the gas but to areas of the atmosphere over the Polar Regions where levels have temporarily declined (although not to zero) during the springtime warming period.
9. The main greenhouse gas, carbon dioxide (CO_2), is also released in this way. Because the carbon content of fossil fuels is intrinsically much higher than that of nitrogen and sulfur, larger quantities of CO_2 are released than NO_2 and SO_2; in

contrast however, CO_2 does not have direct health effects at ambient concentrations, even in polluted areas.

10. A summer or photochemical 'smog' is something of a misnomer as there is no smoke or fog involved.

Chapter Four

1. The following statement from the WHO (2014b) is illuminating with respect to the links between exposure to air pollution and incidence of cancer. "A 2013 assessment by WHO's International Agency for Research on Cancer (IARC) concluded that outdoor air pollution is carcinogenic to humans, with the particulate matter component of air pollution most closely associated with increased cancer incidence, especially cancer of the lung. An association also has been observed between outdoor air pollution and increase in cancer of the urinary tract/bladder."
2. A slightly earlier study, also focussing on $PM_{2.5}$, gives a lower (but still significant) estimate of 19,000 attributable deaths (Yim and Barrett, 2012). Regarding the estimates of attributable deaths, see also note 1, Chapter 1.
3. For example, there is likely to be considerable overlap between the mortality effects of $PM_{2.5}$ and NO_2 because they are typically present in the same polluted air masses. Therefore, the estimated mortality effects of each pollutant should not simply be added together to give a total number of attributable deaths.
4. The daily average relates to concentrations measured midnight-to-midnight on a particular calendar day, not on a rolling 24-hour basis (e.g. not the preceding 24 hours finishing at a random hour during the calendar day).
5. The various pollutants have AQSs based on different averaging times, partly to reflect potential exposure periods. Averaging measured concentrations over a longer period recognises that prolonged exposure to elevated levels is likely

to be a greater risk to health than inhalation of such concentrations over a relatively brief period.

6. This is an excellent web resource run by the Environmental Research Group at Kings College, London. It is a mine of information on air quality and is recommended to anyone interested in the subject.

Chapter Five

1. Of the fifty-five AURN urban background monitoring sites, distance to the nearest busy road is specified in the online descriptions for forty sites. The mean and median distances for these forty sites are 198 m and 145 m, respectively. The site descriptions given for the remaining fifteen sites suggest that the nearest busy road is too far distant to merit a mention.
2. In relation to the health risks of PM, the WHO (2013) states: "There is no evidence of a safe level of exposure or a threshold below which no adverse health effects occur." It adds: "As no threshold for PM has been identified below which no damage to health is observed, the recommended values should be regarded as representing acceptable and achievable objectives to minimize health effects."
3. In relation to interim guideline values, the WHO (2014b) states: "In addition to guideline values, the Air Quality Guidelines provide interim targets for concentrations of PM_{10} and $PM_{2.5}$ aimed at promoting a gradual shift from high to lower concentrations. If these interim targets were to be achieved, significant reductions in risks for acute and chronic health effects from air pollution can be expected. Progress towards the guideline values, however, should be the ultimate objective."
4. These percentages are based on the explanations in the final column that: (i) IT-1 carries a 15% greater mortality risk than the AQG and that IT-2 has 6% less risk than IT-1 (so a 9% greater mortality risk than the AQG) and; (ii) that IT-3 has 6%

less risk than IT-2 (so a 3% greater mortality risk than the AQG).

5. EC statement on dissemination of air quality information (EC, 2015): "The [Air Quality] directives require Member States to ensure that up-to-date information on ambient concentrations of the different pollutants is routinely made available to the public as well as to other organisations. This is done by providing information on websites, teletext, in press and also by public displays. The information needs to be updated as appropriate to the averaging periods. The relation to the different limit and target values needs to be clear. When information or alert thresholds are exceeded Member States need to inform the public about the exceedance and the actions that are eventually taken. This obligation is prescribed in detail in the different directives."
6. The BBC weather website does show the current air quality band for individual location forecasts but they do not routinely appear on the BBC's more widely-viewed TV weather forecasts.
7. In particular, DEFRA now has a Twitter feed with constant updates on air quality measurements and forecasts. This can be followed at: https://twitter.com/DefraUKAIR
8. Electric vehicles do not emit exhaust gases at source; however, the energy they use is provided by power stations and unless these are powered by renewable energy then pollution is still produced at a different location, usually far away from the most polluted urban centres.
9. Printed estimates of the differences in NO_2 emissions from diesel and petrol vehicles cover a wide range; however, calculated emission factors by the European Monitoring and Evaluation Programme and the European Environment Agency (EMEP/EEA, 2013) show that diesel and petrol passenger cars are estimated to emit, respectively, 12.96 and 8.37 g of NO_x (oxides of nitrogen, i.e. nitric oxide and nitrogen dioxide) per kg fuel.

10. Following on from note 9 above, comparative figures for $PM_{2.5}$ are 1.1 and 0.03 g of $PM_{2.5}$ per kg fuel from diesel and petrol passenger cars, respectively (EMEP/EEA, 2013).

Chapter Six

1. There are also plug-in *hybrid* electric vehicles that run partly on internal fuel combustion and partly on a battery charged by electricity.
2. Another theoretical obstacle, the perception that hydrogen-fuelled vehicles pose a dangerous explosion risk, is less serious; in fact, pressurised hydrogen leaking from a damaged fuel tank will quickly dissipate upwards into the atmosphere (hydrogen being much lighter than air) and, as such, is less likely to cause an explosion than petrol leaking onto the ground from a conventional vehicle.
3. Subsidies for bus travel in England in 2014-15 amounted to £2.2 bn, which was also the average annual subsidy for the decade from 2005-06 to 2014-15 (DfT, 2015b); annual bus subsidies in Scotland are approximately £0.3 bn (Transport Scotland, 2015) and in Wales approximately £0.1 bn (Welsh Government, 2013). Rail subsidies for Great Britain are typically around £5 bn per year; total government support for the decade from 2005-05 to 2014-15 averaged, in real terms, £4.98 bn per year (e.g. £4.80 bn in the most recent year, 2014-15, including £1.1 bn for the Crossrail project in southeast England) (ORR, 2015).
4. There are currently more than two hundred LEZs in Europe.
5. Other initiatives introduced in London to tackle emissions from public passenger vehicles have included hybrid buses, age limits for taxis and pollution control upgrades for older buses. These actions are commendable in themselves but there has been much less focus as yet on private vehicles. Considering London's continuing pollution problems and the legal actions relating to poor air quality in London and elsewhere (see Chapter 5), much more remains to be done.

6. Directly in the case of electric vehicles and indirectly in the case of vehicles powered by fuel cells.
7. UK power requirements provided by renewable sources in 2015 (third quarter): total of 23.5%, shared between: onshore wind (5%); offshore wind (4.5%); biomass (9.1%); solar (3.5%); and hydroelectric power (1.4%) (DECC, 2015a).
8. Based on quarterly surveys between 2012 and 2015 of more than 2000 randomly selected adults on each occasion (DECC, 2015b).
9. Biomass combustion is classed as renewable energy because the fuel used – wood and plant material – is sourced from a continuous cycle of plant growth, harvest and regrowth. However, unlike other renewables, air pollutants are released from the combustion process and there are other 'life-cycle' issues to consider, including the transport of the biomass material to the power station and, if biomass is grown specifically for power generation, the use of (energy-intensive) mineral fertilisers.
10. For example, a survey conducted by MORI for the Scottish Executive (Braunholtz, 2003) found that more than half of 1,810 people living close to Scotland's ten largest windfarms were in favour of more turbines being installed. Of those surveyed, 27% said they had thought the landscape would be spoiled before their local windfarm was built but only 12% said this had been borne out in reality.
11. It should be noted that there is no agreed definition of exactly what constitutes an energy subsidy, mainly because of the many and varied support mechanisms that could be considered.
12. One airline, Easyjet, has announced plans to test the use of hydrogen fuel cells to power the engines of an aeroplane whilst it is on the ground.
13. These measures are suggested by Transport and Environment, a consortium of approximately fifty European organisations that are campaigning for sustainable transport. Its website is: http://www.transportenvironment.org.

Chapter Seven

1. In December, 2015, world leaders meeting in Paris agreed pledges to cut greenhouse gas emissions, although these were not legally binding. The pledged cuts, if instigated, would limit global temperature rise to an estimated 2.7 °C, although this is still higher than the 2 °C limit said to be necessary to avoid likely catastrophic impacts. There was also an agreed aim to effectively reach net zero emissions of greenhouse gases by 2050-2100.

Appendix Two

1. The adjustments in RRs due to variations in air pollution across the country yielded estimates ranging from 2-3% excess mortality in rural areas to >8% in parts of London.

References

Barratt, B., Carslaw, D., Fuller, G., Green, D. and Tremper, A. 2012. Evaluation of the impact of dust suppressant application on ambient PM_{10} concentrations in London. Report prepared for Transport for London by the Environmental Research Group of Kings College London. Available at: http://content.tfl.gov.uk/evaluation-dust-suppressants-pmconcentrations.pdf

Bauer, S.E., Tsigaridis, K. and Miller, R. 2016. Significant atmospheric pollution caused by world food cultivation. *Geophysical Research Letters*, **43**, 5394-5400.

Bell, M.L. and Davis, D.L. 2001. Reassessment of the lethal London fog of 1952: Novel indicators of acute and chronic consequences of acute exposure to air pollution. *Environmental Health Perspectives*, **109** (Supplement 3), 389-394.

Brandt, J., Silver, J.D., Christensen, J.H., Andersen, M.S., Bonlokke, J.H., Sigsgaard, T., Geels, C., Gross, A., Hansen, A.B., Hansen, K.M., Hedegaard, G.B., Kaas, E. and Frohn, L.M. 2011. Assessment of Health-Cost Externalities of Air Pollution at the National Level using the EVA Model System, CEEH Scientific Report No 3, Centre for Energy, Environment and Health Report series. Available at: http://www.ceeh.dk/CEEH_Reports/Report_3/CEEH_Scientific_Report3.pdf

Braunholtz, S. 2003. Public Attitudes to Windfarms: A Survey of Local Residents in Scotland. MORI Scotland and Scottish Executive, Edinburgh.

Brimblecombe, P. 1987. The Big Smoke: A History of Air Pollution in London since Medieval Times. Methuen, London.

COMEAP (Committee on the Medical Effects of Air Pollutants). 2010. The Mortality Effects of Long-Term Exposure to Particulate Air Pollution in the United Kingdom. Report produced by the Health Protection Agency, Chilton.

COMEAP (Committee on the Medical Effects of Air Pollutants). 2012. Statement on estimating the mortality burden of particulate air pollution at the local level. August, 2012. Available at: http://webarchive.nationalarchives.gov.uk/20140505104658/http:/www.comeap.org.uk/images/stories/Documents/Statements/FINAL_Local_mortality_burden_statement_August_2012.pdf

Conservatives. 2015. The Conservative Party Manifesto 2015. Available at: https://www.conservatives.com/manifesto

Cook, L.M. and Turner, J.R.G. 2008. Decline of melanism in two British moths: spatial, temporal and inter-specific variation. *Heredity*, **101**, 483-9.

DECC (Department of Energy and Climate Change). 2015a. Energy Trends Section 6: Renewables. Available at: https://www.gov.uk/government/statistics/energy-trends-section-6-renewables

DECC (Department of Energy and Climate Change). 2015b. DECC Public Attitudes Tracker. Available at: https://www.gov.uk/government/collections/public-attitudes-tracking-survey

DECC (Department of Energy and Climate Change). 2015c. Press release: Controlling the cost of renewable energy. Available at: https://www.gov.uk/government/news/controlling-the-cost-of-renewable-energy

DEFRA (Department for Environment, Food and Rural Affairs). 2014a. Air Pollution in the UK 2013. DEFRA, London.

DEFRA (Department for Environment, Food and Rural Affairs). 2014ba. Updated projections for nitrogen dioxide (NO_2) compliance. Available at:
http://uk-
air.defra.gov.uk/assets/documents/no2ten/140708_N02_projection_
tables_FINAL.pdf

DEFRA (Department for Environment, Food and Rural Affairs). 2015a. Air Pollution in the UK 2014. DEFRA, London.

DEFRA (Department for Environment, Food and Rural Affairs). 2015b. Draft plans to improve air quality in the UK: Tackling nitrogen dioxide in our towns and cities. DEFRA, London.

DEFRA (Department for Environment, Food and Rural Affairs). 2015c. Improving Air Quality in the UK: Tackling Nitrogen Dioxide in our Towns and Cities. DEFRA, London.

DEFRA (Department for Environment, Food and Rural Affairs). 2016a. Data Archive. Available at:
http://uk-air.defra.gov.uk/data/

DEFRA (Department for Environment, Food and Rural Affairs). 2016b. Data Archive: Annual and Exceedance Statistics. Available at:
http://uk-air.defra.gov.uk/data/exceedence

DfT (Department for Transport). 2014. Licensed Vehicles 2014. Online factsheet. Available at:
https://www.gov.uk/government/uploads/system/uploads/attachment_data/file/421192/licensed-vehicles-factsheet-2014.pdf

DfT (Department for Transport). 2015a. Vehicle Licensing Statistics, Quarter 1 (Jan-Mar) 2015. Available at: https://www.gov.uk/government/statistics/vehicle-licensing-statistics-january-to-march-2015

DfT (Department for Transport). 2015b. Annual Bus Statistics, Year Ending March 2015 (data tables). Table BUS0502. Available at: https://www.gov.uk/government/statistics/annual-bus-statistics-year-ending-march-2015

Dockery, D.W., Arden Pope, C., Xu, X., Spengler, J.D., Ware, J.H., Fay, M.E., Ferris Jr, B.G. and Speizer, F.E. 1993. An association between air pollution and mortality in six U.S. cites. *New England Journal of Medicine*, **329**, 1753-59.

EC (European Commission). 2013. Environment: New policy package to clean up Europe's air. Published on the EC Press Release Database. Available at: http://europa.eu/rapid/press-release_IP-13-1274_en.htm

EC (European Commission). 2015. Air quality - public information. Published on the EC 'Environment' website. Available at: http://ec.europa.eu/environment/air/quality/legislation/public_info.htm

EEA (European Environment Agency). 2011. Requirements to Air Quality Monitoring in Europe. EEA, Copenhagen.

EMEP/EEA (European Monitoring and Evaluation Programme / European Environment Agency). 2013. EMEP/EEA Air Pollutant Emission Inventory Guidebook 2013. EEA Technical Report No. 12/2013. European Environment Agency, Luxembourg. Available at: http://www.eea.europa.eu//publications/emep-eea-guidebook-2013

EPUK (Environmental Protection UK). 2015. Air Pollution Laws (web page). Available at:
http://www.environmental-protection.org.uk/committees/air-quality/air-pollution-law-and-policy/air-pollution-laws/

ERGKC (Environmental Research Group, Kings College). 2016a. Data downloads. [Data accessible in CSV format.] Available at:
http://www.londonair.org.uk/london/asp/datadownload.asp

ERGKC (Environmental Research Group, Kings College). 2016b. Pollution episodes. Available at:
http://www.londonair.org.uk/london/asp/publicepisodes.asp

ERGKC (Environmental Research Group, Kings College). 2016c. Nitrogen dioxide limits in London, update. London Air website. Available at:
http://www.londonair.org.uk/london/asp/news.asp?NewsId=NO2LimitsLondon2016

Gowers, A.M., Miller, B.G. and Stedman, J.R. 2014. Estimating local mortality burdens associated with particulate air pollution. Public Health England Report PHE-CRCE-010. Public Health England, Chilton. Available at:
https://www.gov.uk/government/uploads/system/uploads/attachment_data/file/332854/PHE_CRCE_010.pdf

Hasheminassab, S., Daher, N., Saffari, A., Wang, D., Ostro, B.D. and Sioutas, C. 2014. Spatial and temporal variability of source of ambient fine particulate matter ($PM_{2.5}$) in California. *Atmospheric Chemistry and Physics*, **14**, 12085-97.

HCEAC (House of Commons Environmental Audit Committee). 2015. The Airports Commission Report: Carbon Emissions, Air Quality and Noise. The Stationery Office Ltd, London.

Hobbs, A. and Harriss, L. 2013. Peak Car Use in Britain. Briefing document by the Parliamentary Office of Science and Technology (POST) for the Commons Transport Select Committee, November 2013. Available at:
http://www.parliament.uk/documents/commons-committees/transport/POST-briefing-on-peak-car.pdf

Hong, S., Candelone, J.P., Patterson, CC. and Boutron, C.F. 1994. Greenland ice evidence of hemispheric lead pollution two millennia ago by Greek and Roman civilizations. *Science*, **265** (5180), 1841-3.

House of Commons Environmental Audit Committee. 2011. Air Quality: A Follow Up Report. The Stationery Office Ltd., London.

IAEA (International Atomic Energy Agency). 2014. Managing Radioactive Waste. Available at:
www.iaea.org/publications/factsheets

IARC (International Agency for Research on Cancer, Working Group on the Evaluation of Carcinogenic Risks to Humans). 2014. Diesel and gasoline engine exhausts and some nitroarenes. IARC Monographs on the Evaluation of Carcinogenic Risks to Humans, Volume 105. IARC, Lyon. Available at:
http://monographs.iarc.fr/ENG/Monographs/vol105/mono105.pdf

IEA (International Energy Agency). 2013. Redrawing the Energy-Climate Map. IEA, Paris. Available at:
http://www.iea.org/publications/freepublications/publication/weo_special_report_2013_redrawing_the_energy_climate_map.pdf

IMF (International Monetary Fund). 2015. Counting the cost of energy subsidies. IMF Survey Magazine, July 17 2015. Available (with related links) at:
http://www.imf.org/external/pubs/ft/survey/so/2015/new070215a.htm

Maher, B.A., Ahmed, I.A.M., Karloukovski, V., MacLaren, D.A., Foulds, P.G., Allsop, D., Mann, D.M.A., Torres-Jardon, R. and Calderon-Garciduenas, L. 2016. Magnetite pollution nanoparticles in the human brain. Proceedings of the National Academy of Sciences of the United States of America, published ahead of print. September 6, 2016, doi: 10.1073/pnas.1605941113.

Nykvist, B. and Nilsson, M. 2015. Rapidly falling costs of battery packs for electric vehicles. *Nature Climate Change*, **5**, 329-32.

ONS (Office for National Statistics). 2016. Statistical Bulletin: Alcohol Related Deaths in the United Kingdom: Registered in 2014. Available at:
http://www.ons.gov.uk/peoplepopulationandcommunity/healthandsocialcare/causesofdeath/bulletins/alcoholrelateddeathsintheunitedkingdom/previousReleases

ORR (Office of Rail and Road). 2016. Government Support to the Rail Industry. Available at:
http://dataportal.orr.gov.uk/browsereports

Peplow, M. 2011. Chernobyl's Legacy. *Nature*, **471**, 562-565.

Raes, F and Seinfeld, J.H. 2009. New directions: Climate change and air pollution abatement: A bumpy road. *Atmospheric Environment*, **43** (32), 5132-5133.

Royal College of Physicians. 2016. Every breath we take: the lifelong impact of air pollution. London, RCP.

SMMT (The Society of Motor Manufacturers and Traders). 2016. December 2015: EV Registrations. Available at:
http://www.smmt.co.uk/2016/01/december-2015-ev-registrations/

Supreme Court. 2015. Judgment: R (on the application of ClientEarth) (Appelant) v Secretary of State for the Environment, Food and Rural Affairs (Respondent). Available at: https://www.supremecourt.uk/decided-cases/docs/UKSC_2012_0179_Judgment.pdf

Tait, C. 2015. Transition by Consent : Meeting Britain's Energy Needs Together. Fabian Society, London.

TfL (Transport for London). 2015. More than 50 all-electric buses to enter service in London. Press release, 15 July 2015. Available at: https://tfl.gov.uk/info-for/media/press-releases/2015/july/more-than-50-all-electric-buses-to-enter-service-in-london

Transport Scotland. 2015. Scottish Transport Statistics No.33 – Datasets. Available at: http://www.transportscotland.gov.uk/statistics/scottish-transport-statistics-no-33-datasets-6495

USDE (United States Department of Energy). 2006. Hydrogen Fuel Cells. DoE Hydrogen Program. Available at: http://www.hydrogen.energy.gov/pdfs/doe_fuelcell_factsheet.pdf

Walton, H., Dajnak, D., Beevers, S., Williams, M. Watkiss, P. and Hunt, S. 2015. Understanding the health impacts of air pollution in London. Report for Transport for London and the Greater London Authority. Available at: http://www.scribd.com/doc/271641490/King-s-College-London-report-on-mortality-burden-of-NO2-and-PM2-5-in-London

Welsh Government. 2013. Written Statement – Bus Services in Wales. Available at: http://gov.wales/about/cabinet/cabinetstatements/2013/busservicesinwales/?lang=en

REFERENCES

WHO (World Health Organisation). 2000. Air Quality Guidelines for Europe, 2nd Edition. WHO Regional Publications, Regional Series, No.91. WHO Regional Office for Europe, Copenhagen. Available at: http://www.euro.who.int/__data/assets/pdf_file/0005/74732/E71922.pdf

WHO (World Health Organisation). 2006. Air Quality Guidelines: Global Update 2005. WHO Regional Office for Europe, Copenhagen. Available at:
http://www.euro.who.int/__data/assets/pdf_file/0005/78638/E90038.pdf

WHO (World Health Organisation). 2009. Country Profiles of Environmental Burden of Disease. WHO, Geneva. Available at: http://www.who.int/quantifying_ehimpacts/national/countryprofile/unitedkingdom.pdf

WHO (World Health Organisation). 2012. Tackling the global clean air challenge. Press release. Available at:
http://www.who.int/mediacentre/news/releases/2011/air_pollution_20110926/en/

WHO (World Health Organisation). 2013. Health Effects of Particulate Matter: Policy Implications for Countries in Eastern Europe, Caucasus and Central Asia. WHO Regional Office for Europe, Copenhagen. Available at:
http://www.euro.who.int/__data/assets/pdf_file/0006/189051/Health-effects-of-particulate-matter-final-Eng.pdf

WHO (World Health Organisation). 2014a. Burden of disease from ambient air pollution for 2012: Summary of results. WHO, Geneva. Available at:
http://www.who.int/phe/health_topics/outdoorair/databases/AAP_BoD_results_March2014.pdf?ua=1

WHO (World Health Organisation). 2014b. Ambient (Outdoor) Air Quality and Health. Factsheet No. 313. Available at: http://www.who.int/mediacentre/factsheets/fs313/en/

WHO (World Health Organisation). 2015. Air pollution costs European economies US$ 1.6 trillion a year in diseases and deaths, new WHO study says. Press release. Available at: http://www.euro.who.int/en/media-centre/sections/press-releases/2015/air-pollution-costs-european-economies-us$-1.6-trillion-a-year-in-diseases-and-deaths,-new-who-study-says

World Bank Group and IHME (Institute for Health Metrics and Evaluation). 2016. The Cost of Air Pollution: Strengthening the Economic Case for Action. International Bank for Reconstruction and Development / The World Bank, Washington DC.

Yim, S.H.L. and Barrett, S.R.H. 2012. Public health impacts of combustion emissions in the United Kingdom. *Environmental Science and Technology*, **46**(8), 4291-4296.

Recommended Websites

It has been argued in this book that the real changes needed to combat air pollution will require serious political action on issues such as public transport and renewable energy. Such change is likely to come about only if the majority of the population (i.e. the electorate) show they want such change, via their statements and actions. Some of the entries in the following list relate to campaign groups that do tremendous work in publicising air quality problems to the wider population and advocating change; they deserve to be more widely known.

Clean Air in London
Campaign group with a mission to achieve full compliance with World Health Organisation guidelines for air quality throughout London and elsewhere.
www.cleanair.london

Environmental Protection UK
A national charity that provides expert advice on air quality.
www.environmental-protection.org.uk

Healthy Air Campaign
News and information from the Healthy Air Campaign – a national coalition of health, transport and environmental NGOs coordinated by ClientEarth to raise awareness and advocate more action by government to tackle air pollution.
www.healthyair.org.uk

Local Air Quality Management
Resources for local authorities including, for example, emissions factors toolkit, background maps and NO_2 calculator.
http://laqm.defra.gov.uk

London Air
Run by the Environmental Research Group at Kings College London. A mine of information about air quality in the UK's most polluted city.
www.londonair.org.uk

National Atmospheric Emissions Inventory
Impressive searchable database with emissions estimates of numerous air pollutants and their sources, going back to 1990.
http://naei.defra.gov.uk

Network for Clean Air
Forms networks with people and communities to improve air quality and runs a citizen science programme for communities to measure air pollution in their neighbourhood.
www.cleanairuk.org (Twitter: @cleanairuk)

Pollutionwatch
Monthly feature in The Guardian newspaper with interesting and topical air pollution news.
www.theguardian.com/environment/series/pollutionwatch

UK-AIR.
A comprehensive website provided by the UK Department for Environment, Food and Rural Affairs. Contains useful information, government publications and reports and the UK's air quality data archive. The archive contains updated pollutant concentrations from the national AURN monitoring network. You can use this to get in-depth information on the levels of air pollutants in your area.
http://uk-air.defra.gov.uk

Index

Page numbers in **bold** refer to tables.

Aberdeen 54

acid rain 10-11, 98n4

aeroplanes *see* air pollution, sources of: air travel

agriculture *see* air pollution, sources of: agriculture

air pollution as a cause of excess mortality 1, 2, 5, 9, 13, 18, 26-28, 43, 44, **45**, 83, 86-87, 89, 92-94, 95n1/n2, 99n2

air pollution episodes 4, 22-25, 28, 33-34, **35**, **36**, 44-45, 59, 73, 81

air pollution, effects on plants and materials 98n4

air pollution, health effects of, 15-16, 17, 19-20, 21, 43, 44, **45**, 86-87, 99n5, 100n2; *see also* air pollution as a cause of excess mortality

air pollution, history of 6-11

air pollution, sources of: agriculture 13, 15, **35**, 79-80, 82, 98n5; air travel 82-83; domestic fires 9, 13, 15, **35**, 80-81; industry 4, 5, 7, 9-11, 13, 14, 17, **35**, 81-82; power generation 13, 17, 63, 74-75, 79; shipping 13, 21, 83; vehicles 12-13, 14, 17, 21, **35**, 38-39, 51-53, 59, 66, **67**, 73, 82, 101n9/n10; waste incineration 13, 15, 82

air quality forecasts 46, 49

air quality index 46, **47**, **48**, 49-50

Air Quality Management Areas (AQMAs) *see* local authorities

air quality objectives 40-42

air quality standards (AQS) **30**, **43**, **45**, **47**; breaches of 31, 32, 40, 41, 42, **91** air travel *see* air pollution, sources of: air travel

airports *see* air pollution, sources of: air travel

ammonia 79, 98n5

ammonium nitrate 16, 98n5

asthma 16, 17, 21, 29, **48**,

Athens 73

attributable deaths, meaning of 92-94, 95n1, 99n3; *see also* air pollution as a cause of excess mortality

Automatic Urban and Rural Network (AURN) 29, 32, 37-42, 100n1

Bath 54

behavioural change 28, 43, 56-57, 59, 67-71, 82, 85, 87

Beijing 73

Belfast 32, 54

benzene 12, **14**, 21

benzo(a)pyrene 22

best available techniques (BAT) 82

biofuel 62-63

biomass combustion 75, 103n7/n9

Birmingham 32, 54, 72

Black Country, the 7

Bogota 73

brain, effects of air pollution on 16

brake wear 15, **67**

Brighton 73

bronchial constriction 16, 17, 21, 29

bronchitis *see* bronchial constriction

buses 12, 39, 54, 61, 69, 70, 72-73, 102n3/n5

1,3-butadiene 22

Caerphilly 54

calcium magnesium acetate (CMA) *see* dust suppressants

cancer *see* carcinogens and carcinogenicity

car bans *see* car rationing

car rationing 73

car sharing / pooling 69

carbon dioxide *see* greenhouse gases

carbon monoxide 21

carcinogens and carcinogenicity 16, 21, 22, 52, 99n1; *see also* lung cancer

cardiovascular and cardiopulmonary disease 16, 26, 43, **45**, **47**, 93

catalytic converter 51, 60

children, exposure to and health effects of air pollution in 2, 9, 16, 17-18, 20, 21, 26, 28, 40, **48**, 85

chronic obstructive pulmonary disease (COPD) 16, 21, 26

classic smog *see* winter smog

Clean Air Acts 9, 20

Clean Air Programme for Europe 58

Clean Air Zones 54

ClientEarth 54

climate change *see* greenhouse gases

coal *see* fossil fuels

Committee on the Medical Effects of Air Pollutants (COMEAP) 2, 27, 46, 92-94

congestion charge 72

Convention on Long-Range Transboundary Air Pollution (CLRTAP) 58
COPD *see* chronic obstructive pulmonary disease (COPD)
countryside *see* rural areas, air pollution in
cultural causes of air pollution and changes required *see* behavioural change
cycling 68-69
deaths from air pollution *see* air pollution as a cause of excess mortality
defeat device 52-53
Delhi 73
Department of Environment, Food and Rural Affairs (DEFRA) 17, 34, 39, 46, 101n7
Derby 54
desert dust 3, 4, 15, **35**, 46
desulfurisation 21, 79
diesel 14, 51-53, 60-61, **72**,
diesel particulate filter 61
domestic fires *see* air pollution, sources of: domestic fires
Dumfries 32
Durham 72
dust suppressants 84
economic costs and benefits (of air pollution and its management) 51, 57, 70, 76
economic tools 52, 55, 61-62, 64, 71-72
Edinburgh 7, 72
education 68
elderly people, effects of air pollution on 9, 20, 29, 85

electric vehicles 63-66, 69, 72, 73, 86, 101n8, 102n1
emissions standards 72-73; for HGVs 73
emissions tests 53
emphysema 16, 29
energy efficiency 55
energy efficiency 78-79, 86
episodes *see* air pollution episodes
Euro emissions standards *see* emissions standards
European Union (EU) 14, 19, 21, 38, 40, **43**, 44, 51, 53, 57, 101n5; UK exit from 38, 97n3
excess mortality *see* air pollution as a cause of excess mortality
exercising in polluted air, health effects of 20, 28, **47**,
farming *see* air pollution, sources of: agriculture
flue gas desulfurisation 79
fossil fuels: combustion of 16-17, 20, 55, 58, 63, 74, 80-81, 83; divestment from 74; *see also* air pollution, sources of
fracking *see* shale gas
freight 73
gas *see* fossil fuels
Glasgow 32, 54
global warming 1, 52, 55, 58, 86, 88-89, 104n1; *see also* greenhouse gases
government *see* political action and inaction
Great Smog of London (1952) 9, 20, 25, 86
'green walls' 84
greenhouse gases 1, 52, 55, 58, 62, 65, 74, 76 88-89, 98n9, 104n1
haulage 73

Hayle 7

heart disease and heart attack *see* cardiovascular disease

Heathrow Airport 83

heavy goods vehicles (HGVs) 54, 61, 72, 73

hybrid vehicles 61, 69, 86, 102n1/n5

hydraulic fracturing *see* shale gas

hydrogen fuel cell *see* hydrogen-powered vehicles

hydrogen-powered vehicles 64-66, 86, 102n2, 103n12

indoor air pollution 95n2

Industrial Revolution 4, 7-8

industry *see* air pollution, sources of: industry

Integrated Pollution Prevention and Control (IPPC) 81-82

internet 67, 71

lead 21-22

Leeds 32, 54

LEZs *see* low emission zones

liquefied petroleum gas (LPG) 62

Local Air Quality Management (LAQM) *see* local authorities

local authorities 32, 39, 50-51

London 2, 6-9, 20, 27, 31, 32, **35**, 42, 51, 54, 69, 72-73, 82-83, 84, 96n1, 102n5, 104n1

London-type smog *see* winter smog

long-range transport of air pollutants 13, 19, 34, **35**, 57-58, 83, 98n6

lorries *see* heavy goods vehicles (HGVs)

Los Angeles 23, 98n5

Los Angeles-type smog *see* photochemical smog

low emission zones (LEZs) 72-73, 102n4

low NOx burners 79

lung cancer 16, 26, 43, **45**, 52, 93, 99n1

lung function, impaired 16, 17-18, 19, 21

Manchester 72

Mexico City 73

Milan 73

monitoring and modelling of air pollutants 29, 32, 33, 37-40, 42

mortality *see* air pollution as a cause of mortality

National Air Quality Strategy 56

nitric oxide 18, 19, 98n7, 101n9

nitrogen dioxide 5, 12, **14**, 16-18, 27, **30**, **31**, 32, **35**, **36**, 40, **41**, 42, **43**, **47**, 51, 60-61, **72**, 79, 80, 83, **91**, 101n9

Norway 86

Norwich 73

notification of poor air quality *see* publicity about air pollution

Nottingham 6, 54, 73

nuclear power 63-64, 77, **78**, 86

oil *see* fossil fuels

Oxford 54, 73

ozone 3, **14**, 18-20, 28, **30**, **31**, 32-33, **35**, **36**, 40, **41**, **43**, **47**, **91**; ozone hole 98n8; ozone layer 18; stratospheric ozone 18

Paris 73

particulate matter (PM) 5, 12, **14**, 14-16, 21, 27, **30**, **31**, 32, **35**, **36**, 40, **41**, **43**, 44, **45**, **47**, 49, 51-52, 60-61, **72**, 79, 80, 83, 84, 88, **91**, 92, 93, 97n2, 98n5, 99n1, 100n2, 102n10 pedestrians *see* walking

Peppered Moth (*Biston betularia*) 10

photochemical smog 18, 23-24, 25, 34, **35**, 99n10

Plymouth 38-39

PM, $PM_{2.5}$, PM_{10} *see* particulate matter

political action and inaction 43, 51, 53-55, 56, 61-62, 76, 77, 78, 85, 86

pollution control technology 79, 82

pollution episodes *see* air pollution episodes

polycyclic aromatic hydrocarbons (PAHs) 16

power stations and power generation *see* air pollution, sources of: power generation

primary pollutants 18

Public Health Acts 8, 9

public transport 59, 69-71, 85, 102n3

publicity about air pollution 44, 46-47, 49-50, 57, 58, 85, 101n5/n6

pulmonary disease *see* chronic obstructive pulmonary disease (COPD)

rail transport 68, 70, 73

renewable energy 55, 63-64, 66, 75-76, 78, 86, 103n7/n10

risk groups 28-29

road transport *see* air pollution, sources of: vehicles

roadside planting 84

rural areas, air pollution in 1, 19, 25, 28, **29**, 32-33, 42, 104n1

Saharan dust *see* desert dust

San Francisco Bay Area 81

Santiago 73

Sao Paulo 73

sea spray 3

secondary pollutants 18

shale gas 55, 74, 76

Sheffield 10, 97n6

shipping *see* air pollution, sources of: shipping

smog episodes *see* pollution episodes

smoke control areas 10, 80-81

solar power *see* renewable energy

Southampton 54

'Spare the Air' scheme 81

speed limits 62

stroke *see* cardiovascular disease

subsidies: fossil fuels 76, 103n11; public transport 70, 102n3; renewable energy 55, 76

sulfur dioxide 5, **14**, 20-21, **30**, **47**, 79, 83

summer smog *see* photochemical smog

technological approaches and solutions 58, 59, 60-61, 63-66, 75, 79, 84

temperature inversions 24-25, 34, 81

Tesla 64

tidal power *see* renewable energy

Toyota Mirai 65

trains *see* rail transport

transboundary pollution *see* long-range transport of air pollutants

tree planting 84

tyre wear 14, 66, **67**

ultra-low emission zones (ULEZs) 73

vehicle bans *see* car rationing

vehicles *see* air pollution, sources of: vehicles

visibility 98n4

volcanoes 3, 4

Volkswagen 52-53, 61

walking 68-69, 85

waste incineration *see* air pollution, sources of: waste incineration

wind power *see* renewable energy

winter smog 24-25, 81; *see also* Great Smog of London, temperature inversions

woodburners *see* air pollution, sources of: domestic fires

World Health Organisation (WHO) 20, 26, 95n2; WHO Air Quality Guidelines 30, 31, 40, 43-44, **45**, 100n3

zero emission vehicles 66, 69, 73, 85; *see also* electric vehicles, hydrogen-powered vehicles

28630659R00079

Printed in Great Britain
by Amazon